Pitman Research Notes in Mathematics Series

Submission of proposals for consideration

Suggestions for publication, in the form of outlines and representative samples, are invited by the Editorial Board for assessment. Intending authors should approach one of the main editors or another member of the Editorial Board, citing the relevant AMS subject classifications. Alternatively, outlines may be sent directly to the publisher's offices. Refereeing is by members of the board and other mathematical authorities in the topic concerned, throughout the world.

Preparation of accepted manuscripts

On acceptance of a proposal, the publisher will supply full instructions for the preparation of manuscripts in a form suitable for direct photo-lithographic reproduction. Specially printed grid sheets can be provided and a contribution is offered by the publisher towards the cost of typing. Word processor output, subject to the publisher's approval, is also acceptable.

Illustrations should be prepared by the authors, ready for direct reproduction without further improvement. The use of hand-drawn symbols should be avoided wherever possible, in order to maintain maximum clarity of the text.

The publisher will be pleased to give any guidance necessary during the preparation of a typescript, and will be happy to answer any queries.

Important note

In order to avoid later retyping, intending authors are strongly urged not to begin final preparation of a typescript before receiving the publisher's guidelines. In this way it is hoped to preserve the uniform appearance of the series.

Longman Group Ltd
Longman House
Burnt Mill
Harlow, Essex, CM20 2JE
UK
(Telephone (0) 1279 426721)

Titles in this series. A full list is available from the publisher on request.

Khalid Alhumaizi

King Saud University, Saudi Arabia

and

Rutherford Aris

University of Minnesota, USA

Surveying a dynamical system: a study of the Gray–Scott reaction in a two-phase reactor

LONGMAN

Longman Group Limited
Longman House, Burnt Mill, Harlow
Essex CM20 2JE, England
and Associated Companies throughout the world.

© Longman Group Limited 1995

First published 1995

ISSN 0269-3674

ISBN 0 582 24688 1

British Library Cataloguing in Publication Data

A catalogue record for this book is
available from the British Library

Library of Congress Cataloging-in-Publication Data

A catalog record for this book is available

Printed and bound in Great Britain
by Biddles Ltd, Guildford and King's Lynn

CONTENTS

List of Figures

LIST OF TABLES

PREFACE

This book is intended for those students of differentiable dynamical systems who like to visualize their systems as vividly and comprehensively as possible and for workers in other areas, such as control theory, who wish to find a simple example on which to try out some new insight or technique. It provides a simple case of three ordinary differential equations with a cubic nonlinearity and six parameters. The model is based on the Gray-Scott autocatalytic reaction A + 2B → 3B, B → C taking place in a two phase reactor, one compartment of this is a reactor and the other a reservoir and the two communicate through a semi-permeable barrier. The two variant forms arise according as A(B) or B(A) is fed to the reactor (reservoir). Almost all the different phenomena that are possible from pitch forks to homoclinics and from period doubling to trancritical Hopf bifrucations occur in the system and there may well be more to be found lurking in the recesses of a six dimensional parameter space.

A secondary theme is the problem of mapping out and presenting the variety of behaviour that a system of this kind and this has led us to use a great number of figures. In the hierarchy of presentation there are 84 presentations of solutions, either by three-dimensional views or by projections onto two-dimensional planes, 141 bifurcation diagrams in which a norm of the solution is plotted against a distinguished parameter (θ, the residence time), 70 branch set diagrams, planes of the two chemical parameters, κ and β, divided into subsets for which the bifurcation diagrams are equivalent, 58 Poincaré, first return maps, and diagrams of other features. Almost all have numerical values attached to all the parameters and are accessible by the use of the index at the back and the list of figures in the front.

The authors are indebted to Martin Golubitsky and his publisher, Springer Verlag, for permission to adapt several figures from Golubitsky and Schaeffer's *Singularities and*

Groups in Bifurcation Theory (1985) and to Peter Gray and Steve Scott and their publisher, Clarendon Press, for permission to use two figures from *Chemical Oscillations and Instabilities* (1990).

The latter stages of the production of so complicated a manuscript would have been intolerable but for the excellence of Jody Peper's word processing. It is too little appreciated that the good reputation of a university department depends, not only on the industry of its students and faculty, but equally upon the ability and labour of the supporting staff. That we have long enjoyed the highest standards in this support and a free and fruitful interchange with our colleagues at all levels in the Department of Chemical Engineering and Materials Science at the University of Minnesota is something the authors have experienced to the full and are both very thankful for. *Floreat Minnesota.!*

K.A. & R.A.

Chapter 1

INTRODUCTION

Le silence éternel de ces espaces infinis m'effraie.

What Pascal faced, spiritually in the depths of his own soul, evokes a purely secular echo in a problem that mathematical modellers face, intellectually in the recesses of their minds. For, though the ability to handle, and even visualize, higher dimensional space is a characteristic of the mathematical mind, the very vastness of a space of (say) six dimensions frustrates the modeller's wish to present a comprehensive picture of the system he is studying. The best he can hope for, in general, is to be able to give complete descriptions of certain cross-sections of lower dimension. These may be entirely satisfactory for some purposes, but, in recent times, the extraordinary complexity of the solutions of disarmingly simple equations has made us aware of the strange behaviour that may lurk in the folds of parameter space.

In the last twenty years a tremendous amount of experience of dynamical systems has accumulated and it would be a scholarly enterprise of the magnitude of the *Corpus Vitrearum Medii Aevi* to organize the existing knowledge of dynamical systems that is scattered through the literature. The stimuli of the Belousov-Zhabotinsky reaction, of the Lorentz equation, of cold flames, of population and morphogenetic studies (to mention but a handful) that have generated an already vast literature are as quick in the community of mathematics today as corresponding stimuli were active in the society that built wayside chapel and Gothic cathedral. Though the reduction of the corpus of dynamical systems to some sort of systematic structure on a scale that befits the variety and interest of its members will probably not be achieved soon, and certainly will not be attempted here, there is a certain structure to the way the examples occur in the better texts. Here it would be invidious to mention too many names without undertaking a comprehensive and critical review, but there can be few who have explored the field without benefiting a great deal from the texts of Wiggins [1988, 1990], Guckenheimer and Holmes [1983] or Golubitsky and Schaeffer [1985, 1988 (the latter with I. Stewart)].

There are a number of monographic treatments of individual systems, such as Tyson on the Belousov-Zhabotinskii reaction [1976] and Sparrow on the Lorenz equations [1982] and it is in this category that this book falls. The Gray-Scott reaction scheme in a stirred tank is among the simplest systems that show a rich variety of behaviour. It has a state space of two dimensions and a parameter space of three and an absolutely comprehensive picture of its behaviour should be possible. Only a sketch of this will be given here for it well covered in Gray and Scott's "Chemical Oscillations and Instabilities" and in the references given later. Our main purpose is to extend the one-phase stirred tank model to the two-phase reactor. This consists of two well-mixed chambers, one in which the reaction takes place, and therefore called the 'reactor', and one which is merely a 'reservoir'. The two are separated by a semi-permeable membrane which allows the transfer of material in the reservoir. Two models are considered in parallel corresponding to the two possibilities of feeding the autocatalytic component to the reactor and the other reactant to the reservoir,or vice-versa. In either case we obtain a three dimensional state-space and six dimensionless parameters.

It is impossible to be comprehensive in six dimensions and our concern lies with the way in which one should go about surveying the system so as to gain as complete an understanding of the system as possible. Obviously we should start with the study of where the steady states lie, and here we get some help from the fact that one of the parameters does not affect the steady state in Model I. Next we use singularity theory to recognize the type of singularity that we have to deal with. Then the stability of the steady states must be discovered and the bifurcation conditions and their degerneracies studied to form a picture of the types of dynamical behaviour. Finally, the breakdown of regular dynamics into chaotic is exemplified, though we make no claim to have discovered more than a sample of the regions where chaos reigns.

1.1. The Gray-Scott cubic autocatalator

It is generally conceded that in order to have 'interesting' behaviour a reaction scheme must have an autocatalytic element. That is, some product of the reaction must tend to make it go faster. This is in contrast to the situation near thermodynamic equilibrium where concentration and temperature differences are small and the system is essentially linear. This contrast probably accounts for phrases such as "far-from-eqilibrium dynamics" and "nonequilibrium transitions" that occur in the chemical literature. The reaction scheme, A + 2B $\rightarrow$ 3B, B $\rightarrow$ C, which has its roots in the work of Salnikov [1948, 1949], was pioneered by Gray and Scott [1983, 1984] as perhaps the simplest such reaction. Moreover they allowed it to take place in a stirred tank in preference to imposing pool hypotheses. This system will be referred to as the Gray and Scott CSTR and it is this that we generalize here to a two-phase reactor. The equations of the Gray and Scott CSTR, which will be derived in the next chapter, are

$$dx/d\tau = (1-x)/\theta - xy^2 \qquad (1.1)$$

$$dy/d\tau = (\beta-y)/\theta + xy^2 - \kappa y \qquad (1.2)$$

They are not structurally stable as they stand, but Merkin, Needham and Scott [1987] and Balakotaiah [1987] have shown how to make them so: the former by adding an uncatalysed $A \rightarrow B$; the latter by making both reactions reversible. It has been shown that the even simpler $A + B \rightarrow 2B$ is not sufficiently non-linear to give the full range of possible behaviour and the objection that an apparently termolecular reaction is involved can be answered by showing that it is the limit of two bimolecular steps [Gray, Scott and Aris, 1988].

There is no need to give a separate survey of the Gray and Scott CSTR for this is to be found in Gray and Scott's book [1990]. Some features appear later in this work as a limiting case of the two-phase reactor (see Chapter 12 and references). Aris [1990] gave a brief comparison of some of these reaction schemes and there is some relevant material in the second part of his "Introduction to the mathematical background of chemical reactor analysis" [1986]. But we will consider it from the point of view of our method and the manner of presentation of our results.

In the presentation of any dynamical system what we would ultimately like to do is to give a gallery of all possible solutions and associate them with the appropriate subsets of parameter space. As we have said above, this can only be done for systems of small dimensions. The Brusselator, as the reaction scheme devised by Prigogine and Nicholis and much used by the Brussels school is often called, has two state variables and two parameters. It can be thought of as the irreversible reaction $A + B = C + D$ which however goes by the steps $A \rightarrow X, B + X \rightarrow Y + D, Y + 2X \rightarrow 3X, X \rightarrow C$. The core of the mechanism is clearly the cubic autocatalysis of the third step, for, in its simplest form, A and B are held constant in what is called the 'pool hypothesis' i.e., that they are in great excess and scarcely depleted during the course of the reaction. The resulting equations, which will only be quoted here, are:

$$dx/d\tau = \alpha - \beta x + x^2 y - x$$

$$dy/d\tau = \beta x - x^2 y$$

where x, and y, the dimensionless concentrations of X and Y, are the two state variables and α and β, the dimensionless levels of the A and B pools. This system has a unique steady state at $x = \alpha$, $y = \beta/\alpha$ and its Jacobian has eigenvalues of

$$\tfrac{1}{2}\left[\beta - 1 - \alpha^2 \pm \left\{ (\beta - (1 - \alpha)^2)(\beta - (1 + \alpha)) \right\}^{1/2}\right]$$

Thus the unique steady state is stable if $\beta < 1 + \alpha^2$, is a focus if β is between $(1 - \alpha)^2$ and $(1 + \alpha)^2$ and a node outwith these limits. In fact it undergoes a Hopf bifurcation as the parameter point (α, β) crosses the line $\beta = 1 + \alpha^2$ at any point, so that there is a limit cycle for parameter values corresponding to any point above this line.

Thus the total behaviour of the Brusselator can be displayed in a tableau as in Figure 1.1. The central panel is the α, β-plane with the Hopf line and the two node-focus transition lines. Six parameter points are marked with the letters A to F and the corresponding phase planes surround the central panel. Two of the parameter points, B and D, are on the Hopf curve and two, E and F, have one of the parameters set equal to zero. In addition the two marked C' and C" show that making the core reactions of the mechanism reversible, by replacing their reaction rates βx and $x^2 y$ by $\beta(x - \kappa y)$ and $x^2(y-k'x)$, respectively, tends to destroy the limit cycle. These two phase portraits are of an enlarged model, namely that of the Brusselator with a reversible core, in which the core reactions must be regarded as kinetically independant in spite of their being stoicheiometrically identical under the pool hypothesis. This enlarged model has a four

5

dimensional parameter space and a comprehensive exploration of this is already a much more hazardous enterprise. It illustrates the vastness of higher dimensional parameter space referred to above and the way one may have to take sections of lower dimensions. We shall not explore this example further but return to the Gray-Scott model which is both simpler and more realistic.

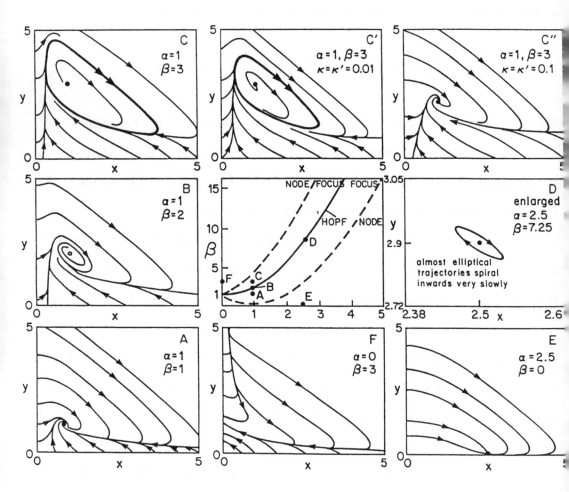

Figure 1.1. The behaviour of the Brusselator displayed. See text for description.

As may be seen from Equations 1.1 and 1.2 the Gray-Scott stirred tank has two state variables, x and y, and three parameters, θ, β and κ, so it is not impossible to think of being truly comprehensive and divide the three dimensional parameter space up into regions such that the dynamics are qualitatively the same for two sets of parameter values that correspond to points in the same region. By the dynamics being the same we mean that the phase planes representing the solution are topologically equivalent. This implies that the two have the same number of steady states (critical points) and that these are of the same character, and the same number of periodic solutions of the same disposition. Being a two-dimensional system it cannot exibit anything more exotic than a periodic solution. Equivalent regions in parameter space can be descibed by their boundaries, so it is natural to look for the transitions or bifurcations where there are changes in the number or stability of the steady states. These can be exhibited clearly in bifurcation diagrams in which one of the parameters is singled out and a 'norm' of the solution is plotted against it. Though the parameter space of the Gray-Scott CSTR is only three dimensional it is so finely reticulated in places that it would be necessary to present it in two-dimensional sections. In this case the same effect is produced by singling out θ, the dimensionless reaction time (or Damköhler number as it is known to chemical engineers), as the distinguished parameter. The norm of the solution is taken as the conversion $z = 1 - x$ for the steady states and the greatest and least values of z for a periodic solution. When we come to the two-phase equations z will be a reservoir concentration and we shall use x as the norm.

1.2. Multiplicity of steady states.

The steady states of the dynamical system governed by equations (1.1) and (1.2) are easy to find by setting the derivatives equal to zero, eliminating y and expressing everything in terms of z. This gives a cubic equation

7

$$[(1 + \theta\kappa)^2 / \theta] \, z = (1 - z)(z + \beta)^2 \qquad (1.3)$$

for z. An admissible value of z will lie between 0 and 1 and will be at the intersection of the straight line through the origin of slope $[(1 + \theta\kappa)^2 / \theta]$, which is what the left hand side of the equation is and the cubic curve on the righthand side. A very pretty separation of parameters is obtained, for the cubic depends only on β and at that rather simply. If no tangent of this cubic curve goes through the origin, there can be no more than one intersection of any line through the origin. Since the tangent to the cubic at its point of inflection does not go through the origin if $\beta > 1/4$, this is a sufficient condition for the uniqueness of the steady state. For $\beta < 1/4$, there are two tangents of the cubic (let their slopes be $4\sigma(\beta)$ and $4\Sigma(\beta)$) that go through the origin (Figure 1.2 in which f denotes either side of the Equation 1.3), and if the slope of the left hand side of the equation lies, between these values there are three intersections and so three steady states. But it is easy to see that the slope of the line on the left hand side of Equation 1.3 comes down from the

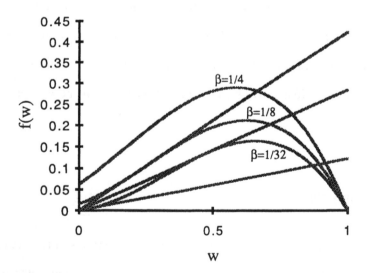

Figure 1.2. The cubic curve in the steady state analysis of the Gray-Scott CSTR.

vertical when $\theta = 0$, to a minimum of 4κ when $\theta = 1/\kappa$, and goes back to the vertical as θ gets larger and larger. If κ, the least slope of the left side, is greater than $\Sigma(\beta)$, the greatest slope of an original tangent on the right side, there can never be more than one intersection, and that at a small value of z. Hence the bifurcation diagram is as in Figure 1.3a. If κ lies in the interval $\sigma < \kappa < \Sigma$, there is a single interval on the θ axis for which there are three steady states but the two are born remote from the one that is there all the time and they die without ever having connected up with it. This creates the isola shown in Figure 1.3b. Finally, if $\kappa < \sigma$ the 'mushroom' structure, in which there are two regions of triplicity (Figure 1.3c), obtains.

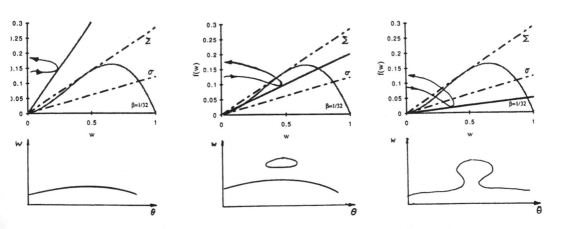

Figure 1.3 Possible intersections of straight line and cubic curve of Equation (1.3).

We have given the geometrical argument because it is so simple and direct, but the systematic way of doing this is to use singularity theory. The master work on this is Golubitsky and Schaeffer's two volumes [1985, 1988] the second with I. Stewart., but there are many others which will be referred to later.

Multiplicity considerations come first, but they are not sufficient to define the dynamics, since several qualitatively different phase planes may be consistent with any given constellation of steady states. However, because it is so simple and serves as an introduction to the way in which results are presented later, we show in Figure 1.4 thebranch set diagram for multiplicity. It divides the positive quadrant of κ, β-space into

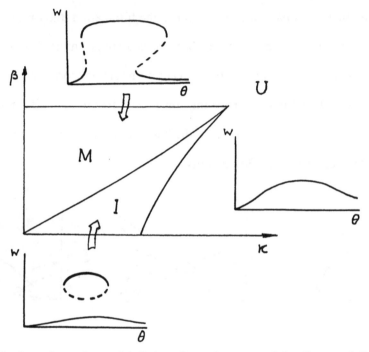

Figure 1.4. The branch set for multiplicity of steady states of the Gray and Scott CSTR.

three regions: U for the unique solution for all values of θ; I for the isola which gives only one region of triplicity; and M that gives two. The transitions take place on the lines of this branch set at values of θ for which there are vertical tangents.

1.3. Stability of the steady state. Hopf bifurcation.

Nothing has yet been said about the stability of the steady states. This is determined by the eigenvalues of the Jacobian at the steady state which must both lie in

10

the left half-plane if the steady state is to be stable. Like the multiplicity, the stability of any one is a function of the parameters, and we think of all as flowing with the distinguished parameter, θ, the remaining parameters, κ and β remaining constant. If the eigenvalues are real, we have a node; if complex conjugate, a focus. A node can lose its stability if the larger of its eigenvalues slips into the right half-plane through the origin; it then becomes a saddle. If, however, a pair of complex conjugates passes cleanly into the right-half plane across the imaginary axis a stable limit cycle is shed around the now unstable steady state. A lengthy analysis, due to Poincaré's pioneering work and Hopf's 1942 paper, distinguishes between the supercritical Hopf transition in which a stable limit cycle is shed by a focus in the act of becoming unstable and a subcritical transition which sees the death of an unstable limit cycle which had previously existed within a stable limit cycle. By passing 'cleanly' is meant that they go across alone and with a finite speed, and as before we are interested in the degeneracies when these conditions are not satisfied for these again mark the boundaries between regions of qualitatively similar behaviour.

The bifurcation diagram (the graph of a solution norm versus the distinguished parameter) can be enhanced to show the periodic solutions. It can be 'read' in the sense that with each of the subintervals of the θ axis that does not contain a transition we can associate a typical phase plane, so to speak, a 'letter' in the 'name' of the bifurcation diagram. This bifurcation diagram will have been generated for a particular pair of values of the other two parameters, κ and β. But a particular bifurcation diagram may be associated with more than one point in κ, β-space and its 'name' may thus serve as a 'label' for a curve or subset of κ, β-space. The κ, β-space, divided into equivalent regions by transitional curves is called the "branch set" diagram.

For the Gray and Scott CSTR the full branch set is given in their monograph on "Chemical Oscillations and Instabilities" [Gray and Scott, 1990; Figures. 8.13 and 14]. and is reproduced here with permission as Figures 1.5 to 1.7. They show that there are

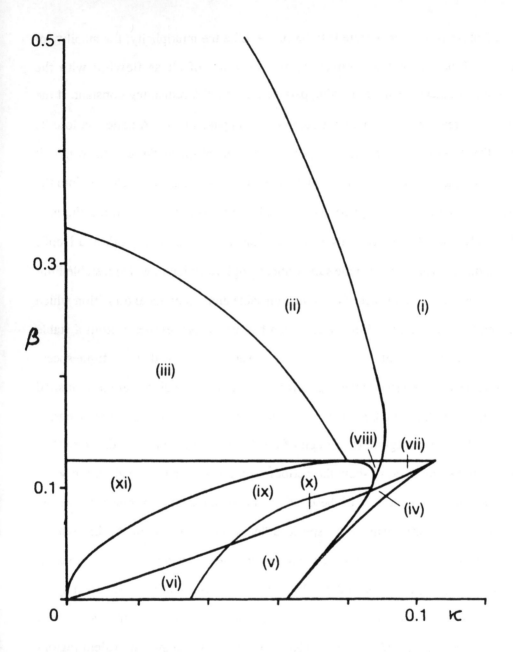

Figure 1.5. The division of the β, κ-parameter region into 11 regions by the various loci of stationary-state and Hopf bifurcation degeneracies. The qualitative forms of the bifurcation diagrams for each region are given by (i) - (xi) in Figure 1.6, (based on Figure 8.13 of Gray and Scott [1990] by permission of Oxford University Press).

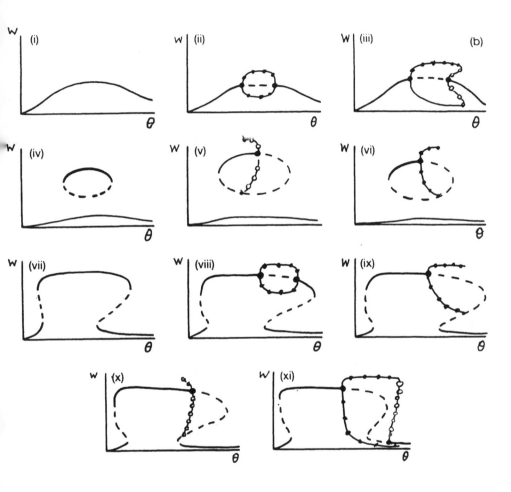

Figure 1.6. The eleven bifurcation diagrams. In this figure solid lines (circles) represent stable stationary states (limit cycles) and broken curves (open circles) correspond to unstable states (limit cycles) (based on Figure 8.13 of Gray and Scott [1990] by permission of Oxford University Press).

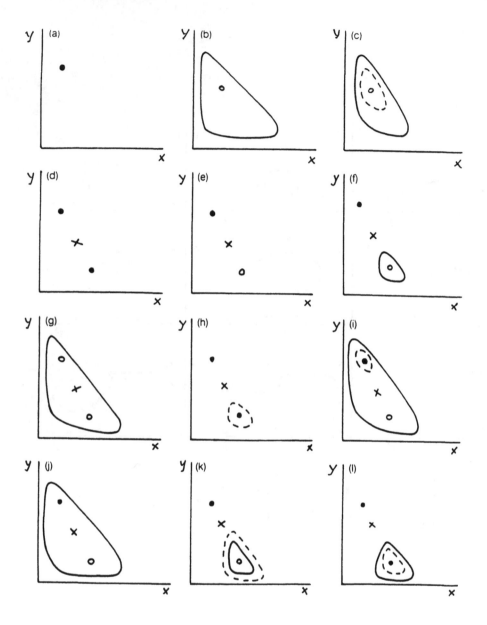

Figure 1.7. The different phase plane portraits identified for cubic autocatalysis with decay; (based on Figure 8.14 of Gray and Scott [1990] by permission of Oxford University Press).

14

twelve distinct phase portraits which they can associate with open subsets of the κ, β-space. Since the system has only three parameters this gives a complete and very satisfying portrait of the system.

1.4. Stability of periodic solutions. Chaos.

Just as the stability of steady states may change and generate periodic solutions, so may the stability of a periodic solution break down in a variety of different ways. In two dimensions this is limited to a Hopf or homoclinic birth or death or to the coalescence of a stable and unstable limit cycle in the periodic equivalent of a saddle-node transition. But in three dimensions there is a greater variety including period doubling, when the closed curve of a periodic solution becomes the two edges of an infinitesimally narrow Möbius strip, bifurcation to a torus as well as homoclinic maneuvers. These transitions are also the gateways to chaos and raise the question of how this may be distinguished from a periodic solution of a very large period. All these possibilities are opened up by the extension of the reactor to a two-phase reactor-reservoir, for the reservoir concentration requires a third equation..This means that the phase plane of the Gray and Scott CSTR is replaced by a three dimensional phase space, though the solutions are sometimes represented by their projections on the planes $x = 0$, $y = 0$, and $z = 0$.

1.5. How to use this book.

This book may be regarded as an ecological study of a particular species (the Gray-Scott autocatalator) in a particular habitat (the two-phase reactor). The species comes in two varieties (depending on which species is fed to which phase) each described by three ordinary differential equations with six parameters. These parameters are divided into three groups:

- θ, a measure of the intensity of the reaction (Damköhler number);

- κ, β-chemical parameters, (ratio of the side reaction rate to that of the main reaction and the ratio of the concentrations of A and B fed to the reactor, respectively);

- σ, λ, ρ-physical parameters, (ratio of flow rates to the two phases, dimensionless mass transfer resistance between phases, and ratio of the volumes of the two phases, respectively).

The distinguished parameter, in the sense of bifurcation theory, is θ.

The behaviour is presented as curves in the state space (coordinates x, y, z) parametrized by the independent variable, time; or by the projections of such solution curves on the coordinate planes; or by showing x as a function of time. These phase portraits are referred to as 'model trajectories' (MT). A chosen norm of the solution plotted against the distingished parameter is a bifurcation diagram (BD) and any section of this for constant θ is a summary of an MT which can be unpacked from the conventions of the BD. A branch set diagram (BS) is in the space of the chemical parameters κ and β and is a plane divided into regions such that the BDs of all points in the same region are topologically identical. Finally in our case a BS can be atached to a point in the residuary parameter space of σ, λ, ρ. Presumably one ought to be able to carve up the residuary space into regions of qualitatively identical brsanh set diagrams. It is here that we have succumbed to the overpowering vastness of high dimensional space and far from illuminating whole regions of this space we have but made short stabs into the darkness by varying one parameter at a time. Features of a BS have occasionally been plotted against a residuary parameter (e.g., Figure 7.18).

16

An understanding of the structure of the subject and the organization of the book will allow readers to find their ways to whatever they seek. The short chapters on theory are not intended to be a textbook, but are brief expositions of the routes we have followed and, of course, give the nomenclature. After setting up the equations, the layout of the material is as follows.

Subject	Theory	Model I	Model II
Establishing equations		Ch. 2.2	Ch. 2.3
Steady state analysis	Ch. 3	Ch. 4	Ch. 5
Dynamics	Ch. 6	Ch. 7	Ch. 8
Chaos	Ch. 9	Ch. 10	Ch. 11

An index to all the numerate figures is provided and is arranged lexically in ascending order of the residuary parameters, σ, λ, ρ. It can be used with the complete lists of legends that follows the Table of Contents and gives ready access to the parameter values that will give a desired behaviour.

Chapter 2

MATHEMATICAL MODELLING

In this part we formulate the mathematical model of two isothermal well-stirred tanks or cells separated by a semi-permeable membrane. The system is based on a cubic autocatalytic reaction scheme. The autocatalytic reaction consists of two steps:

Autocatalytic reaction:

$$A + 2B \rightarrow 3B \qquad rate = k_1AB^2$$

Decay reaction:

$$B \rightarrow C \qquad rate = k_2B$$

The reaction occurs only in one tank, the "reactor", to which either the reactant A or the autocatalytic species B is fed directly. The other component diffuses through the membrane from the other compartment which we will call the "reservoir. Two models are considered here; in the first model, it is the reactant A that is fed through the membrane, while in the second it is the autocatalyst B that enters the reactor through the membrane. For both cases a linear relationship between the flux from reservoir to reactor and their concentration difference is assumed.

2.1   Cubic Autocatalysis in a CSTR

A necessary requirement for any chemical reaction system that is going to show multiplicity or oscillation is that it should have a "feedback mechanism". Some intermediate species or product of the process must be able to influence the rate of earlier steps. This effect may be thermal, as in the classic case of a first-order exothermic reaction in a CSTR, where the temperature rise influences the reaction rate constant. The feedback may be chemical, for example as in "autocatalysis" in which at least one product of a reaction increases the reaction rate and thus its own rate of production. An early study by Lin [1979] showed that general autocatalytic reactions of the form

$$mA + nB \rightarrow (m + n)B, \qquad rate = k\, A^m B^n$$

in a CSTR gave multiple steady states. The behaviour of the models is determined by a single ordinary differential equation and the models are known as 'one variable systems'. Therefore oscillatory response for such models with one reaction step is not possible.

A number of other autocatalytic models involving more than one dependent variable have been thoroughly studied [Boissonade, 1976; Gray and Scott, 1990a]. Gray and Scott [Gray and Scott, 1983] have studied a general kinetic scheme

$$A + 2B \xrightarrow{k_1} 3B, \quad rate = k_1 AB^2$$

$$B \xrightarrow{k_2} C, \quad rate = k_2 B$$

The rate law for the autocatalytic reaction, rate = $k_1 AB^2$, involves the third power of concentration and therefore the reaction is known as "cubic autocatalysis". Cubic autocatalytic systems are considered as the easiest and simplest to analyze and understand. The mathematical model for this scheme in an isothermal CSTR (Figure

19

2.1) consists of two mass-balance equations, one for the reactant and the other for the autocatalyst, given by :

$$V \frac{dA}{dt} - Q(A_0 - A) - Vk_1AB^2$$

$$V \frac{dB}{dt} - Q(B_0 - B) - Vk_1AB^2 - Vk_2B \tag{2.1}$$

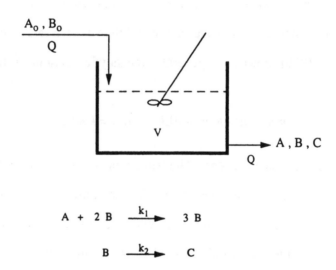

$$A + 2B \xrightarrow{k_1} 3B$$

$$B \xrightarrow{k_2} C$$

Figure 2.1. Cubic autocatalytic reactions in a CSTR

where A_o and B_o are the concentrations of the reactant and the autocatalyst, respectively, in the inflow stream, V is the volume of the reactor, Q is the volumetric flow rate to the reactor, k_1 is the rate constant of the autocatalytic reaction and k_2 is the rate constant for the decay reaction. To reduce the system parameters, a dimensionless form has been derived:

$$\frac{dx}{dr} = \frac{1-x}{\theta} - xy^2$$

$$\frac{dy}{dr} = \frac{\beta - y}{\theta} - xy^2 - \kappa y \qquad (2.2)$$

where $x = A/A_o$, $y = B/B_o$, and $\beta = B_o/A_o$ are the dimensionless concentrations, $\tau = k_1 A_o^2 t$ is the dimensionless time, $\theta = k_1 A_o^2 V/Q$ is the dimensionless residence time and $\kappa = k_2/k_1 A_o^2$ is the dimensionless decay-rate constant. Gray and Scott [Gray and Scott, 1983; 1984] have shown that this system can exhibit a wide variety of behaviour. They showed that more than one steady state could occur and that the steady states were often unstable; oscillatory behaviour was also a distinct possibility. It has been shown that this isothermal system has many resemblances to the first-order exothermic reaction system and appears to display similar patterns of behaviour [Gray and Scott 1990b].

2.2   Model I: Two-Tank System Coupled Through the Reactant A

A schematic representation of this case is shown in Figure 2.2. The reactant A enters the reactor from the reservoir through the membrane while the autocatalyst B is fed directly to the reactor. Both reaction steps take place only in the reactor. A linear mass-transfer relationship is assumed. First we define the state variables of this model. A is the reactant concentration and B is the autocatalyst concentration in the reactor. The third state variable A^* is the concentration of the reactant in the reservoir.   The governing equations for the concentrations A, B, and A^* are:

$$V \frac{dA}{dt} = k_cS\,(A^* - A) - QA - Vk_1AB^2$$

$$V \frac{dB}{dt} = Q\,(B_o - B) + Vk_1AB^2 - Vk_2B$$

$$V \frac{dA^*}{dt} = -k_cS\,(A^* - A) + qA_o - qA^* \tag{2.3}$$

where A_o and B_o are the inlet concentrations for the reactant and the autocatalyst, respectively; V and v are the volumes of the reactor and tank, respectively; Q and q are the volumetric flow rates to the reactor and the tank, respectively; k_1 is the rate constant for the autocatalytic reaction, k_2 s the rate constant for the decay reaction, k_c is the mass transfer coefficient through the membrane and S is the membrane surface area.   This

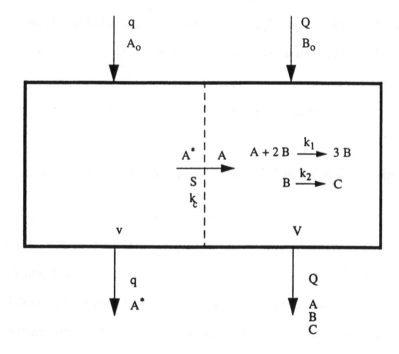

Figure 2.2. First model: coupled through the reactant

model appears to offer ten parameters that might be varied. To make this model more attractive we need to reduce the system parameters by recasting the model in dimensionless form. Actually, the qualitative behaviour of the model may not be sensitive to the absolute values of these parameters but rather to a smaller number of quotients formed from them.

Define the dimensionless concentrations :

$$x = \frac{A}{A_{ref}}, \quad y = \frac{B}{A_{ref}}, \quad z = \frac{A^*}{A_{ref}}, \quad \beta = \frac{B_o}{A_{ref}}$$

The reference concentration, A_{ref}, will be defined later. The dimensionless time can be formed using the chemical time t_{ch} formed from the autocatalytic reaction rate constant k_1. Thus if we multiply k_1 by the square of the reference concentration and then take the reciprocal, we have:

$$t_{ch} = \frac{1}{k_1 A_{ref}^2}$$

A dimensionless time τ and a dimensionless decay rate constant κ are defined as:

$$\tau = \frac{t}{t_{ch}} = k_1 A_{ref}^2 t, \quad \kappa = k_2 t_{ch} = \frac{k_2}{k_1 A_{ref}^2}$$

The dimensionless form of the mean residence time of the reactor, given by the volume of the reactor V divided by the volumetric flow rate Q, is defined as:

$$\theta = \frac{V}{Q t_{ch}} = \frac{V k_1 A_{ref}^2}{Q}$$

23

Dividing the model equations by $V k_1 A_{ref}^2$, and using A_o as a reference concentration and the model dimensionless forms , the model equations become:

$$\frac{dx}{dt} = \frac{z - x}{\lambda} - \frac{x}{\theta} - xy^2$$

$$\frac{dy}{dt} = \frac{\beta - y}{\theta} + xy^2 - \kappa y$$

$$\rho \frac{dz}{dt} = \frac{z - x}{\lambda} + \frac{\sigma}{\theta}(1 - z) \qquad (2.4)$$

where $\lambda = V k_1 A_o^2 S$ is the reciprocal of the dimensionless mass transfer coefficient (mass transfer resistance), $\rho = v/V$ is the ratio of the tank volumes and $\sigma = q/Q$ is the ratio of the flow rates. Obtaining the model in this form, we have reduced the system parameters to six: $\theta, \beta, \kappa, \sigma, and \rho$. As a natural constraint, all these parameters must be positive.

2.2.1 Special Cases

The first limiting case can be obtained when $\lambda \to \infty$ or, alternatively, when the flow rate ratio and tank volume ratio approach zero, i.e., $\sigma, \rho \to 0,0$. The rates of change of the reactor state, x and y in our model, given by equation 2.4, can now be expressed as:

$$\frac{dx}{dt} = \frac{-x}{\theta} - xy^2$$

$$\frac{dy}{d\tau} = \frac{\beta^* - y}{\theta} + xy^2 - \kappa y \qquad (2.5)$$

This case is equivalent to the Gray and Scott case with zero initial reactant concentration $(A_O = 0)$.

The second special case is found by observing that with mass transfer resistance parameter approaching zero, $\lambda \to 0$. The concentrations of component A in the reactor and in the tank now become equal. In this case the system is not affected by the mass transfer rate; therefore we expect the system to behave like a single CSTR with two feed inlets.

The third case is obtained based on the desire to reduce the dimensionless equations to the single CSTR case (Gray and Scott model). We modify our model by using the following definition for the reference concentration:

$$A_{ref} = A_O \frac{\sigma}{\sigma + 1}$$

With this definition the model becomes:

$$\frac{d\hat{x}}{d\hat{\tau}} = \frac{\hat{z} - \hat{x}}{\hat{\lambda}} - \frac{\hat{x}}{\hat{\theta}} - \hat{x}\hat{y}^2$$

$$\frac{d\hat{y}}{d\hat{\tau}} = \frac{\hat{\beta} - \hat{y}}{\hat{\theta}} + \hat{x}\hat{y}^2 - \hat{k}\hat{y}$$

$$\rho\frac{d\hat{z}}{d\hat{\tau}} = -\frac{\hat{z} - \hat{x}}{\hat{\lambda}} + \frac{1}{\hat{\theta}}(\sigma + 1 - \sigma\hat{z}) \qquad\qquad (2.6)$$

where the state variables and the system parameters ($\hat{\theta}, \hat{\beta}, \hat{k}, \hat{\lambda}$) are defined as in the last section, using this new reference concentration. For very small values of ρ, the left side of the third equation of (equation 2.6) vanishes and the concentrations of the reactant in the tank and the reactor are related as:

$$\hat{z} = \frac{\hat{x}\hat{\theta} + \hat{\lambda}(\sigma + 1)}{\hat{\theta} + \sigma\hat{\lambda}}$$

As the flow rate ratio, σ, vanishes, the previous relationship becomes:

$$\hat{z} = \frac{\hat{x}\hat{\theta} + \hat{\lambda}}{\hat{\theta}}$$

Substituting $\hat{z}$ in terms of $\hat{x}$ in the first equation of this model, equation 2.6, leads us to a two variable model, namely a cubic autocatalytic reaction in a single CSTR (Gray and Scott Case):

$$\frac{d\hat{x}}{d\hat{t}} = \frac{1 - \hat{x}}{\hat{\theta}} - \hat{x}\hat{y}^2$$

$$\frac{d\hat{y}}{d\hat{t}} = \frac{\hat{\beta} - \hat{y}}{\hat{\theta}} + \hat{x}\hat{y}^2 - \hat{\kappa}\hat{y} \qquad (2.7)$$

Now we need to show that the reference concentration used in defining system parameters for this limiting case is similar to the one used by Gray and Scott. The reference concentration for this special case can be written in terms of the system flow rates:

$$A_{ref} = A_o \frac{q}{q + Q}$$

As σ becomes small, $QA_{ref} = qA_o$ = total feed of A. This form simply implies that A_{ref} should be taken as the total amount of A entering the system divided by the feed flow rate to the reactor.

2.3   Model II: Two-Tank System Coupled Through the Autocatalyst B

A schematic representation of this model is shown in Figure 2.3.   For this case the autocatalyst is fed to the reactor through the membrane.   Note that the decay reaction takes place in both vessels (the reactor and the tank).   A linear mass-transfer relationship is used in this case.   The state variables for this model are: A, the reactant concentration

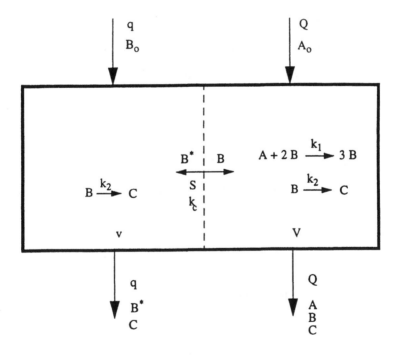

Figure 2.3. Second model: coupled through the autocatalyst

in the reactor, B, the autocatalyst concentration in the reactor and B^*, the autocatalyst concentration in the tank. The mass balance equations for this model are:

$$V\frac{dA}{dt} = Q(A_o - A) - Vk_1AB^2$$

$$V\frac{dB}{dt} = k_cS(B^* - B) - QB + Vk_1AB^2 - Vk_2B$$

$$v\frac{dB^*}{dt} = -k_cS(B^* - B) + q(B_o - B^*) - vk_2B^* \qquad (2.8)$$

where A_o and B_o are the initial concentrations for the reactant and the autocatalyst, respectively; V and v are the volumes of the reactor and the tank, respectively; Q and q are the volumetric flow rates to the reactor and to the tank, respectively; k_1 is the rate constant for the autocatalytic reaction, k_2 is the rate constant for the decay reaction, k_c is the mass transfer rate coefficient through the membrane and S is the membrane surface area. As in the first model, we have ten parameters that can be varied. To simplify this problem, a dimensionless form similar to the one obtained in the first section will be derived. The state variables of the model in dimensionless form are defined as:

$$x = \frac{A}{A_{ref}}, \quad y = \frac{B}{A_{ref}}, \quad z = \frac{B^*}{A_{ref}}, \quad \beta = \frac{B_o}{A_{ref}}$$

The dimensionless time can be obtained using the chemical time t_{ch}, formed from the autocatalytic reaction rate constant k_1. Thus if we multiply k_1 by the square of the reference concentration and then take the reciprocal we have

$$t_{ch} = \frac{1}{k_1A_{ref}^2}$$

28

A dimensionless time τ and a dimensionless decay rate constant κ are defined as

$$\tau = \frac{t}{t_{ch}} = k_1 A_{ref}^2 t, \quad \kappa = k_2 t_{ch} = \frac{k_2}{k_1 A_{ref}^2}$$

The dimensionless form of the mean residence time of the reactor, given by the volume of the reactor V divided by the volumetric flow rate Q, is defined as:

$$\theta = \frac{V}{Q t_{ch}} = \frac{V k_1 A_{ref}^2}{Q}$$

Dividing the model equations by $V k_1 A_{ref}^2$ and using the initial reactant concentration A_o as our reference concentration, the model equations become:

$$\frac{dx}{d\tau} = \frac{1 - x}{\theta} - xy^2$$

$$\frac{dy}{d\tau} = \frac{z - y}{\lambda} - \frac{y}{\theta} + xy^2 - \kappa y$$

$$\rho \frac{dz}{d\tau} = -\frac{z - y}{\lambda} + \frac{\sigma}{\theta}(\beta - z) - \rho \kappa z \tag{2.9}$$

where $\lambda = \dfrac{V k_1 A_o^2}{k_c S}$ is the reciprocal of the dimensionless mass transfer coefficient (mass transfer resistance), $\rho = \dfrac{v}{V}$ is the ratio of the tank volumes and $\sigma = \dfrac{q}{Q}$ is ratio of the flow rates. In this dimensionless model, six parameters appear explicitly: θ, β, κ, λ, σ, and ρ. A natural constraint on our model is that all parameters must be positive.

2.3.1 Special Cases

In the first case we consider the system with very small mass transfer rate, i.e., $\lambda \to \infty$ or very small flow rate ratio and tank volume ratio; i.e., $\sigma, \rho \to 0,0$. The model in this case, equation 2.9, is equivalent to the Gray and Scott case with no autocatalyst inflow :

$$\frac{dx}{d\tau} = \frac{1-x}{\theta} - xy^2$$

$$\frac{dy}{d\tau} = -\frac{y}{\theta} + xy^2 - \kappa y \qquad\qquad (2.10)$$

The second case arises when the mass transfer resistance parameter vanishes $(\lambda \to o)$. The autocatalyst concentrations in both tanks are equal and the system dynamics are dominated by the first equation of this model. We expect that the system would not be affected by mass transfer rate but would behave like a single CSTR.

The third case is derived based on the desire to relate this three variable model as a limiting case to the simple form, of a cubic autocatalytic reaction in a single CSTR; i.e., the Gray and Scott case. We introduce the following parameter:

$$\beta' = \beta \, \frac{\sigma}{\sigma + 1}$$

With this definition, the model equations become:

$$\frac{dx}{d\tau} = \frac{1-x}{\theta} - xy^2$$

$$\frac{dy}{d\tau} = \frac{z-y}{\lambda} - \frac{y}{\theta} + xy^2 - \kappa y$$

$$\rho\frac{dz}{d\tau} = -\frac{z-y}{\lambda} + \frac{1}{\theta}\left(\beta'(\sigma+1) - \sigma z\right) - \rho\kappa z \qquad (2.11)$$

For very small values of the tank volume ratio ρ, the concentrations of the autocatalyst in the reactor, y, and in the tank, z are related as:

$$\frac{(z-y)}{\lambda} = \frac{1}{\theta}\left(\beta'(\sigma+1) - \sigma z\right)$$

When σ vanishes, this relationship becomes:

$$\frac{z-y}{\lambda} = \frac{\beta'}{\theta}$$

Substituting this equality in the second equation of our model, equation 2.11, leads us to the Gray and Scott case form:

$$\frac{dx}{d\tau} = \frac{1-x}{\theta} - xy^2$$

$$\frac{dy}{d\tau} = \frac{\beta'-y}{\theta} + xy^2 - \kappa y \qquad (2.12)$$

31

We will now show that in the limit when $\sigma \rightarrow 0$, $\rho \rightarrow 0$, the state variables and the system parameters have the same definition as those of the single CSTR case. The reference concentration used in our model and in Gray and Scott case is the initial concentration of the reactant; therefore the state variables and the residence time θ and the dimensionless decay rate constant κ are the same in both cases. The quantity β' in the Gray and Scott case is simply the relative inflow concentration of the autocatalyst to the reactant, $\dfrac{B_o}{A_o}$. In our model β' does not have the same meaning because we have these two components entering the system through two flow rates, q and Q. β' is given in terms of the flow rates as:

$$\beta' = \frac{B_o q}{A_o (q + Q)}$$

and as $\sigma \rightarrow 0$, $q + Q \rightarrow Q$ and the parameter β' can be written as:

$$\beta' = q B_o / Q A_o$$

In this expression, the quantity in the numerator is the total amount of B entering the system divided by the by the flow rate to the reactor, which gives the inflow autocatalyst concentration. When further divided by the reactant concentration, we clearly have the ratio of the incoming fluxes of the autocatalyst and the reactant, the same definition of β as that of Gray and Scott.

2.4 Isothermal and Nonisothermal Systems

We have seen that both coupled models are simple extensions of the two-variable cubic autocatalysis in a CSTR (Gray and Scott CSTR). The exchange of the material between the two cells adds a third variable to the model. The second model (coupling through the autocatalyst) is also closely related to the nonisothermal first-order exothermic reaction in CSTR with the extraneous thermal capacitance of a cooling jacket:

$$\frac{dx}{d\tau} = -x + \alpha_1 (1 - x) \exp \left(\frac{y}{1 + y/\alpha_2} \right)$$

$$\frac{dy}{d\tau} = y + \alpha_3 \alpha_1 (1 - x) \exp \left(\frac{y}{1 + y/\alpha_2} \right) - \alpha_4 (y - \alpha_5) - \alpha_4 \alpha_6 (y - z)$$

$$\alpha_7 \frac{dz}{d\tau} = \alpha_4 \alpha_6 (y - z) \tag{2.13}$$

where α_i, $(i = 1, ..., 7)$ are the system parameters defined by Planeaux [Planeaux and Jensen, 1986] and Uppal [Uppal, Ray, and Poore, 1974]. The states of the nonisothermal model, x, y, and z, are the dimensionless concentration of the reactant, the temperature of the CSTR, and the temperature of the jacket, respectively. In this case, the exchange of heat with the jacket, provides a third interacting variable. The nonisothermal model is characterized by the extremely nonlinear behaviour due to the exponential dependence on the reactor temperature (thermal feedback) which results in very stiff differential equations. Thus, the isothermal models are typically easier to examine numerically when compared to the analogous equations for exothermic reactions. The use of the autocatalysis to replace the thermal feedback in the nonisothermal model results in structurally similar sets of equations to the second model (coupling through the autocatalyst). The analysis of the nonisothermal system has revealed that complex dynamics - quasi-periodic, multiple-periodic solution and chaos can occur even in his simple model [Planeaux and Jensen, 1986]. Therefore, we expect our system to display most of the various types of complex behaviour found in the nonisothermal system. At the same time we might quite expect that the subtler structure of the Arrhenius temperature dependence gives opportunity for types of behaviour not possible with the simple power law of autocatalysis.

Chapter 3

STEADY STATE ANALYSIS

The most common phenomena encountered in nonlinear systems is multiplicity of equilibrium points when there are several steady states for the same set of parameters. Our first concern is to devise a global picture of the parameter space and systematically subdivide it so that this multiplicity may be understood. Singularity theory developed by Golubitsky and Schaeffer [Golubitsky and Schaeffer, 1985] provides a framework in which the different kinds of multiplicity can be classified according to conditions which depend on the number of vanishing derivatives of the equilibrium function. This theory has been used by chemical engineers [Balakotaiah and Luss, 1982; 1983; D'Anna, Lignola, and Scott, 1986; Farr and Aris, 1986; Kay, Scott, and Lignola, 1987] since the pioneering work of Balakotaiah and Luss [Balakotaiah and Luss, 1981] to analyze the multiplicity phenomena of the classic problem. In the first section we introduce the concept of static bifurcation. Following the analysis of [Balakotaiah and Luss, 1981; 1982], we present a brief account of singularity theory in the second section.

3.1   Static Bifurcation

A bifurcation is said to take place when there is a change in the number or the stability of solutions of an equation as a parameter varies. Consider the parameterized dynamic system

$$\frac{dx}{dt} = F(x, \theta, \alpha)$$

$$x \in R^n, \theta R^l, \alpha \in R^m \tag{3.1}$$

F is a C^r function on $R^n \times R^l \times R^m$, x are the state variables, θ is the distinguished bifurcation parameter and α is the (constant) vector of the other parameters. Suppose that equation 3.1 has a steady state (x_{ss}, θ_{ss}), the local stability of this solution can be studied by linearizing equation 3.1 about this point. The linearized model is given as

$$\frac{d\zeta}{dt} = D_x F(x_{ss}, \theta_{ss}) \zeta \tag{3.2}$$

where the state variables ζ represent the deviation of the system states from the steady state solution (x_{ss}, θ_{ss}). If the fixed point is hyperbolic, i.e., none of the eigenvalues of $D_x F(x_{ss}, \theta_{ss})$, the Jacobian matrix, lie on the imaginary axis, we know that the stability of (x_{ss}, θ_{ss}) is determined by the linear equation, equation 3.2. Thus, varying θ slightly does not change the nature and the stability of this solution since hyperbolic points are structurally stable. The real problem starts when the steady state points (x_{ss}, θ_{ss}) are not hyperbolic i.e., when the Jacobian matrix has some eigenvalues on the imaginary axis. In this case, perturbing the system around θ_{ss} might lead to a new dynamic behaviour. For example, steady solutions can be created or destroyed or time dependent behaviour can be created.

Static bifurcation, which involves creating or destroying time-independent solutions, is related to the simplest way in which the Jacobian matrix is nonhyperbolic. This case arises when $D_x F(x_{ss}, \theta_{ss})$, has a single zero eigenvalue with the remaining eigenvalues having non-zero real parts. The orbit structure near the static bifurcation point can be determined by studying a single ordinary differential equation:

$$x_1 = F_1(x_1, \theta_1) \tag{3.3}$$

The static bifurcation conditions are

$$F_1(x_{ss}, \theta_{ss}) = 0, \quad \left(\frac{\partial F_1}{\partial x_1}\right)_{x_{ss}, \theta_{ss}} = 0 \tag{3.4}$$

This condition is a necessary, but not sufficient, condition for static bifurcation to occur. Points satisfying equation 3.4 simultaneously are called singularities. Generally speaking, this single equation can be obtained using local bifurcation theories such as the center manifold theory [Guckenheimer and Holmes, 1983; Wiggins, 1990]. The set of all singular points in the (x, θ, α) is called the singular set. The projection of the singular set onto the (θ, α) space is called the branch or catastrophe set. The most common static bifurcation point is the saddle-node bifurcation (limit or turning point). The bifurcation diagram $(x \ vs \ \theta)$ in Figure 3.1 arises by joining two branches of the curves of the solutions: on one branch are locally stable (node) solutions and on the other branch are locally unstable (saddle) solutions. In order for a system to undergo a saddle-node

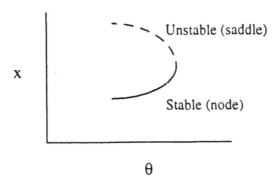

Figure 3.1. Saddle-node bifurcation

bifurcation at (x_{ss}, θ_{ss}), equation 3.4 must be satisfied in addition to the following conditions:

$$(a) \ \frac{\partial F_1}{\partial \theta} \neq 0, \quad (b) \ \frac{\partial^2 F_1}{\partial x^2} \neq 0 \qquad (3.5)$$

The first condition in equation 3.5 implies that a unique curve of steady states passes through the bifurcation point and the second equation implies the curve lies on one side of θ_{ss}. When equation 3.5a does not hold, one of two kinds of static bifurcation could occur: the first one is called transcritical or simple bifurcation (Figure 3.2a). The other is the isola center (Figure 3.2b). When the second condition, equation 3.5b, does not hold, a third type of static bifurcation called hysteresis obtains (Figure 3.2c). The more derivatives of the equation 3.3 that vanish before the first non-zero term, the higher the order of singularities and the more complicated the static bifurcation.

3.2   Singularity Theory

For a wide variety of equations, problems concerning multiple solutions can be reduced to a single scalar function

$$f(x, \theta, \alpha) = 0$$

$$x \in R^l, \theta \in R^l, \alpha \in R^m \qquad (3.6)$$

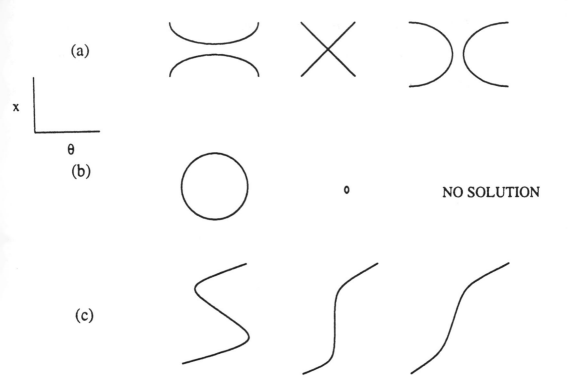

Figure 3.2. (a) Simple bifurcation, (b) Isola center, (c) Hysteresis

The singularity theory deals with this kind of equation. For the moment we assume $\alpha = (\alpha_1, \alpha_2 ..., \alpha_m)$, a vector of auxiliary parameters, is fixed and that the singularity under investigation is situated at the origin is such that

$$f(0,0) = 0, \quad f_x(0,0) = 0 \qquad (3.7)$$

where f is smooth and has continuous derivatives of all orders. A first step toward a classification of singularities is to ignore differences that are not essential.

Definition (1) $f(x, \theta)$ *and* $g(y, \vartheta)$ *are said to be contact equivalent when their qualitative features are the same. Or more formally, there is a scaling factor* $T(y, \vartheta)$ *such that for the transformation*

$$(y, \vartheta) \rightarrow (x(y, \vartheta), \theta(\vartheta))$$

the following holds:

$$g(y, \vartheta) = T(y, \vartheta) f(x(y, \vartheta), \theta(\vartheta)),$$
$$x(0, 0) = 0, \ \theta(0) = 0, T(0, 0) = 0,$$
$$\frac{\partial x(0, 0)}{\partial y} > 0, \quad \frac{\partial \theta(0)}{\partial \vartheta} > 0$$

This definition allows us to identify bifurcation problems that differ only by a change of coordinates. For example the saddle-node bifurcation problem

$$f(x, \theta) = x^2 - \theta = 0$$

and the bifurcation problem

$$g(g, \vartheta) = 1 - \vartheta - \cos(y) = 0$$

are contact equivalent next to the origin, since

$$g(g, \vartheta) = 2f(\sin(y/2), \vartheta/2), \quad and \quad f(x, \theta) = 0.5g(2\sin^{-1}(x), 2\theta)$$

40

Thus the local features of a complex nonlinear equation can be determined by the analysis of the features of much simpler contact equivalent polynomial equations. The next important concept is that of stability of the bifurcation problem. $f(x, \theta)$ is defined to be structurally stable if $f(x, \theta) + h(x, \theta, \varepsilon)$ is contact equivalent to $f(x, \theta)$ in the neighborhood of the origin for all sufficient small smooth functions $h(x, \theta, \varepsilon)$. For example, the normal form of the saddle-node singularity

$$f(x, \theta) = x^2 - \theta = 0$$

is structurally stable in the sense that by perturbing $f(x, \theta)$ to

$$f_1(x, \theta) = x^2 - \theta - \varepsilon x = 0$$

the qualitative features of these bifurcation problems are unchanged for small ε. On the other hand, the bifurcation diagram of the hysteresis normal form

$$f_2(x, \theta) = x^3 - \theta = 0$$

is unstable since the bifurcation diagram of

$$f_3(x, \theta) = x^3 - \theta - \varepsilon x = 0$$

assumes different forms depending on whether ε is positive or negative (Figure 3.3). However each of the bifurcation diagrams of f_3 is stable for any $\varepsilon \neq 0$. Singularity theory provides a method for studying the stability of such bifurcation problems via the unfolding concept.

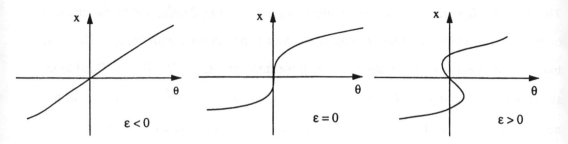

Figure 3.3. Bifurcation diagram for $f_3(x, \theta) = x^3 - \theta - \varepsilon x = 0$.

Definition (2) *An unfolding of $f(x, \theta)$ is an m-parameter family*

$$U(x, \theta, \alpha_1, \alpha_2, ..., \alpha_m)$$

with

$$U(x, \theta, 0, 0, ..., 0) = f(x, \theta)$$

Here $\alpha_1, \alpha_2, ..., \alpha_m$ are additional parameters.

For example, f_3 is an unfolding of f_2. Attaching an additional term is also an unfolding; hence the above definition is still too general to be useful for classification. This leads to another definition:

Definition (3) *A universal unfolding is an unfolding with two features: (1) It includes all possible small perturbation of $f(x, \theta)$ (up to contact equivalence), (2) It uses the minimum number of parameters ($\alpha_1, \alpha_2, ..., \alpha_m$).*

This number m is called the codimension of $f(x, \theta)$. It is the lowest dimension of the parameter space α which is necessary to observe a given bifurcation phenomena. For example

$$f_4 = x^3 + \varepsilon^2 x - \theta = 0$$

is unfolding of f_2, but it is not universal since it can not describe the diagram corresponding to the negative ε in Figure 3.3; while

$$f_5 = x^3 - \varepsilon_1 x^2 - \varepsilon_2 x - \theta = 0$$

is not universal since because the two parameters ε_1, ε_2 are not both necessary. Therefore, only f_3 is universal since it has the minimum number of parameters which are needed to show all perturbed bifurcation forms.

The bifurcation patterns of equation 3.6 are stable for most values of the parameter vector α, for example the bifurcation diagrams of f_3 is stable for all values of ε except zero. But there exists a set of values of α at which the function defined by equation 3.6 has the most degenerate singularity. Perturbing the system around these points results in forming several qualitatively different stable bifurcation diagrams. These singular points are characterized by the vanishing of a finite number of the derivatives of f with respect to x and the bifurcation parameter θ. In addition, there exists always a polynomial function, normal form, which is contact equivalent to f next to these singular points. Here we give a special case of a more general theorem proved by Golubitsky and Schaeffer [1985]. The following theorem provides the normal forms and the universal unfoldings for singularities characterized by vanishing the derivatives of f with respect to x only.

Theorem *Suppose that $f(x, \theta)$ has a singular point $(x, \theta) = (0, 0)$ that obeys*

$$f(0, 0) = 0, \quad \frac{\partial^i f\,(0,0)}{\partial x^i} = 0, \quad i = 1, 2, ..., j$$

$$\zeta = \frac{\partial^{j+1} f(0, 0)}{\partial x^{j+1}} \frac{\partial f(0, 0)}{\partial \theta} \neq 0$$

Then locally, $f(x, \theta)$ is contact equivalent to the normal forms

$$x^{j+1} + \theta, \quad \zeta > 0;$$
$$x^{j+1} - \theta, \quad \zeta < 0$$

The universal unfolding is

$$U(x, \theta, \alpha) = x^{j+1} - \alpha_{j-1} x^{j-1} - ... - \alpha_1 x \pm \theta.$$

This theorem shows that if such a singular point exists, the maximum number of solutions of equation 3.6 next to the singular point is $j + 1$. In the remaining part of this section, we summarize the results for systems which have at maximum three solutions. The normal forms for all singularities of these systems are polynomial functions with cubic or quadratic order in x as shown in Table 3.1. This table shows the conditions for any function f to be contact equivalent to certain normal form. The universal unfoldings of these elementary bifurcations, which are not unique, are shown in Table 3.2. The recognition problem of universal unfoldings for these normal forms is given as following:

Lemma *Let G be an unfolding of a single scalar function g which is equivalent to the normal form h(x, θ), given in Table 3.3. Then G is a universal unfolding if and only if the determinant of the matrix corresponding to this normal form in Table 3.3 is nonzero at the singular point.*

Singularity theory shows that the qualitative features of the bifurcation diagrams of all singularities shown in Table 3.2 change when the parameters α cross one of two surfaces. The first set is the set of all points in the parameter space α satisfying

$$f(x, \theta, \alpha) = \frac{\partial f(x, \theta, \alpha)}{\partial x} = \frac{\partial f(x, \theta, \alpha)}{\partial \theta} = 0$$

This set is called the isola variety *(I)*. Eliminating of x and Q from these three equations gives a single algebraic equation in α, defining a hypersurface. When α crosses the isola variety, two limit points (saddle-node bifurcation points) appear or disappear, so that either one isolated curve disappears (Figure 3.2b) or the bifurcation diagram separates locally into two isolated curves (Figure 3.2a). The second, called the hysteresis variety *H*, is the set of all points in the parameter space satisfying

$$f(x, \theta, \alpha) = \frac{\partial f(x, \theta, \alpha)}{\partial x} = \frac{\partial^2 f(x, \theta, \alpha)}{\partial x^2} = 0$$

When α crosses the hysteresis variety typically two limit points appear or disappear as shown in Figure 3.2c. The perturbed bifurcation diagrams for the pitchfork and winged cusp singularities are given in term of these two varieties in Figure 3.4. Four stable bifurcation patterns are identified around the pitchfork points while seven possible bifurcation diagrams can be found close to the winged cusp singularities.

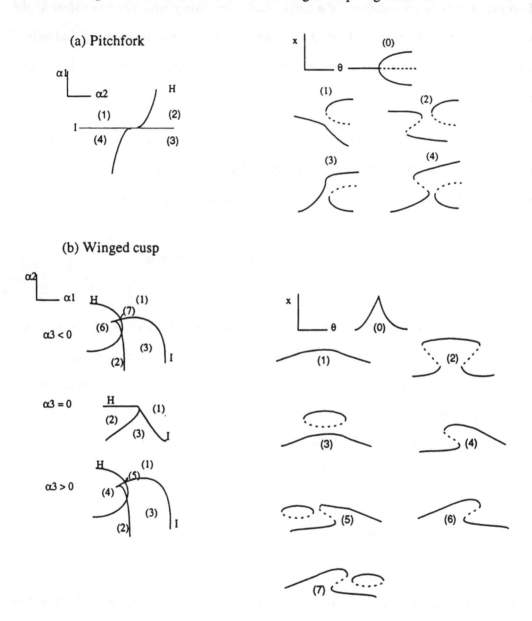

Figure 3.4. Perturbed bifurcation diagrams for the pitchfork and winged cusp problems.

Type	Codimension	Normal Form	Conditions
Saddle-node	0	$\varepsilon x^2 + \delta\theta$	$f = f_x = 0$
Isola Center	1	$\varepsilon(x^2 + \theta^2)$	$f = f_x = f_\theta = 0$
Simple (Mushroom)	1	$\varepsilon(x^2 - \theta^2)$	$f = f_x = f_\theta = 0$
Hysteresis	1	$\varepsilon x^3 + \delta\theta$	$f = f_x = f_{xx} = 0$
Pitchfork	2	$\varepsilon x^3 + \delta\theta x$	$f = f_x = f_{xx} = f_\theta = 0$
Winged cusp	3	$\varepsilon x^3 + \delta\theta^2$	$f = f_x = f_{xx} = f_\theta = f_{x\theta} = 0$

Table 3.1. Conditions and normal forms ($\varepsilon = \pm 1,\ \delta = \pm 1$).

Type	Universal Unfolding	Bifurcation Diagram $\delta = 1, \varepsilon = 1$	Bifurcation Diagram $\delta = -1\ \varepsilon = 1$
Saddle-node	$\varepsilon x^2 + \delta\theta$		
Isola, Simple	$\varepsilon(x^2 + \delta\theta^2 + \alpha_1)$		
Hysteresis	$\varepsilon x^3 + \delta\theta + \alpha_1 x$		
Pitchfork	$\varepsilon x^3 + \delta\theta x + \alpha_1 + \alpha_2 x^2$		
Winged Cusp	$\varepsilon x^3 + \delta\theta^2 + \alpha_1 + \alpha_2 x + \alpha_3 \theta x$		

Table 3.2. Universal unfoldings.

Normal Form $h(x, \theta)$	Matrix
Saddle-node: $\varepsilon x^2 + \delta\theta$	-
Isola: $\varepsilon(x^2 + \theta^2)$	G
Simple: $\varepsilon(x^2 - \theta^2)$	G
Hysteresis: $\varepsilon x^3 + \delta\theta$	$\begin{pmatrix} g_\theta & g_{\theta x} \\ G_{\alpha_1} & G_{\alpha_1 x} \end{pmatrix}$
Pitchfork: $\varepsilon x^3 + \delta\theta x$	$\begin{pmatrix} 0 & 0 & g_{x\theta} & g_{xxx} \\ 0 & g_{x\theta} & g_{\theta\theta} & g_{\theta xx} \\ G_{\alpha_1} & G_{\alpha_1 x} & G_{\alpha_1 \theta} & G_{\alpha_1 xx} \\ G_{\alpha_2} & G_{\alpha_2 x} & G_{\alpha_2 \theta} & G_{\alpha_2 xx} \end{pmatrix}$
Winged Cusp: $\varepsilon x^3 + \delta\theta^2$	$\begin{pmatrix} 0 & 0 & g_{x\theta} & g_{xxx} \\ 0 & g_{x\theta} & g_{\theta\theta} & g_{\theta xx} \\ G_{\alpha_1} & G_{\alpha_1 x} & G_{\alpha_1 \theta} & G_{\alpha_1 xx} \\ G_{\alpha_2} & G_{\alpha_2 x} & G_{\alpha_2 \theta} & G_{\alpha_2 xx} \end{pmatrix}$

Table 3.3.   The recognition problem for universal unfoldings of singularities with codimension less than three.

Chapter 4

STEADY STATE ANALYSIS: MODEL I

The steady state behaviour for the first case of the two-phase reactor with coupling through the reactant, will be examined by means of a single non-linear algebraic equation (Section 4.1). This allows us to use the singularity theory to analyze the steady-state locus qualitatively in Section 4.2.

4.1   Steady State Relationship

The steady-state model has dimensionless concentrations x_{ss}, y_{ss}, z_{ss} satisfying the simultaneous equations

$$0 = \frac{z_{ss} - x_{ss}}{\lambda} - \frac{x_{ss}}{\theta} - x_{ss} y_{ss}^2 \qquad (4.1)$$

$$0 = \frac{\beta - y_{ss}}{\theta} + x_{ss} y_{ss}^2 - \kappa y_{ss} \qquad (4.2)$$

$$0 = \frac{z_{ss} - x_{ss}}{\lambda} + \frac{\sigma}{\theta}(1 - z_{ss}) \qquad (4.3)$$

Equation 4.3 gives a simple linear relationship between the concentration of the reactant A in the two tanks:

$$z_{ss} = \frac{x_{ss}\theta + \sigma\lambda}{\theta + \sigma\lambda} \qquad (4.4)$$

A linear relationship between x_{ss} and y_{ss} is found by substituting equation 4.4 into equation 4.1 and adding equations 4.1 and 4.2:

$$y_{ss} = \frac{\theta\sigma(1 - x_{ss}) - x_{ss}(\theta + \sigma\lambda) + \beta(\theta + \sigma\lambda)}{(\theta + \sigma\lambda)(1 + \kappa\theta)} \tag{4.5}$$

A single steady state equation can be obtained by substituting y_{ss} and z_{ss} from equation 4.5 and equation 4.4 into equation 4.1:

$$F(x_{ss}, \theta, \kappa, \beta, \sigma, \lambda) = \Sigma(\theta, \kappa)\,\Psi(x_{ss}, \theta, \sigma, \lambda) - x_{ss}(\Psi(x_{ss}, \theta, \sigma, \lambda) + \beta)^2 \tag{4.6}$$

where

$$\Sigma(\theta, \kappa) = \frac{(1 + \kappa\theta)^2}{\theta} \quad ,$$

$$\Psi(x_{ss}, \theta, \sigma, \lambda) = -(1 + \xi)x_{ss} + \xi$$

and

$$\xi(\theta, \sigma, \lambda) = \frac{\theta\sigma}{\theta + \sigma\lambda}$$

The states of this system are bounded. The maximum value for x_{ss} is obtained by assuming that A is not consumed by reaction and the concentrations of the reactant A in both tanks are equal $(x_{ss} = z_{ss})$. For this case, mass balance for the component A, in terms of the original parameters is:

$$qA_o = qA_{max} + QA_{max}$$

or

$$x_{ss,max} = \frac{\sigma}{\sigma + 1}$$

The autocatalyst concentration has a maximum value y_{max} when $x = 0$ and $\kappa = 0$. Under these conditions, equation 4.5 becomes:

$$y_{ss} = \frac{\theta\sigma + \beta(\theta q + \sigma\lambda)}{(\theta + \sigma\lambda)}$$

When the mass transfer resistance parameter, λ, tends to zero, y has its maximum value:

$$y_{ss,max} = \sigma + \beta$$

The concentration of the reactant in the reservoir z_{ss} has its maximum value when there is no mass transfer between the reservoir and the reactor. Equation 4.4 gives:

$$z_{ss,max} = 1$$

To summarize, the states of this system are bounded as follows:

$$0 \le x \le \frac{\sigma}{\sigma + 1}, \quad 0 \le y \le \sigma + \beta, \quad 0 \le z \le 1$$

We are interested in the way that the steady state x_{ss} depends on the positive system parameters. For this model, the steady state equation 4.6 does not depend on the volume ratio ρ. The residence time θ, which is the easiest parameter to vary is selected to be the bifurcation parameter. The steady state equation, 4.6, is a cubic in x_{ss}, so for a given value of θ there will be either one or three real positive roots. For this kind of equation, singularity theory defines two types of codimension-one singularities, namely hysteresis and isola-mushroom singularities. Pitchfork (codimension-two) and winged cusp (codimension-three) singularities are also defined for a cubic single scalar equation.

4.2   Singularity Theory

4.2.1   Codimension-one Singularities

(A)   Hysteresis

The conditions for the appearance/disappearance of a hysteresis loop are given as

$$F = F_x = F_{xx} = 0 ;$$

so the steady state condition must be satisfied ($F = 0$) and the first two partial derivatives of F with respect to the steady state solution must vanish. In addition, there is the constraint that a number of other derivatives must remain non-zero, so F_θ, $F_{x\theta}$, F_{xxx} do not equal zero.

For our system the hysteresis conditions are (note that $\Sigma_x = 0$ and $\Psi = 0$)

$$F = \Sigma\Psi - x_{ss}(\Psi + \beta)^2 = 0 \qquad (4.7)$$

$$F_x = \Sigma\Psi_x - (\Psi + \beta)^2 - 2x_{ss}\Psi_x(\Psi + \beta) = 0 \qquad (4.8)$$

$$F_{xx} = -4\Psi_x(\Psi + \beta) - 2x_{ss}\Psi_x^2 = 0 \qquad (4.9)$$

where

$$\Psi_x = -(1 + \xi)$$

Equation 4.9 has two solutions; one of them is physically realizable:

$$(\Psi + \beta) = -\frac{1}{2}x_{ss}\Psi_x$$

Using the definitions of Ψ and Ψ_x, we can express the steady state solution in terms of the system parameters:

$$x_{ss} = \frac{2\xi + \beta}{3\xi + 1}$$

Substituting the expression of Σ from equation 4.7 ($\Sigma = x_{ss}(\Psi + \beta)^2/\Psi$) into equation 4.8 and with some manipulation, equation 4.8 can be written as

$$F_x = (x_{ss}\Psi_x - \Psi)(\Psi + \beta) - 2x_{ss}\Psi_x\Psi = 0$$

which can be used to define the parameter β in term of ξ by using the definitions of x_{ss}, Ψ, and Ψ_x defined above:

$$\beta = \frac{\xi}{8} \qquad (4.10)$$

The reactant concentration at the hysteresis singularity can be expressed as a function of θ, σ, and λ:

$$x_{ss} = \frac{3}{4}\frac{\xi}{\xi + 1} \tag{4.11}$$

The dimensionless decay rate constant can be found by substituting the values of β and x from equation 4.10 and equation 4.11 into the steady state condition $F = 0$:

$$\Sigma = \frac{(1 + \kappa\theta)^2}{\theta} = \left(\frac{3}{4}\right)^3\frac{\xi^2}{\xi + 1} \tag{4.12}$$

Here the quantity ξ plays a parametric role in defining equations 4.10, 4.11, and 4.12. These equations define the hysteresis singularity. The hysteresis boundaries can be defined as a surface in the four parameter κ, β, σ, λ- space:

$$\kappa = \sqrt{\frac{3^3}{4}\frac{8\beta(\sigma - 8\beta)}{\sigma\lambda(8\beta + 1)} - \frac{\sigma - 8\beta}{8\beta\sigma\lambda}} \tag{4.13}$$

The hysteresis singularities for our system are shown in Figure 4.1. This figure shows the κ, β region for $\sigma = 1$ and $\lambda = 1$ at which hysteresis bifurcation diagrams can be found. The other derivatives at these conditions should remain non-zero. $F_{xxx} \neq 0$ can be proven analytically for the hysteresis singularities:

$$F_{xxx} = -6\Psi_x^2 \neq 0$$

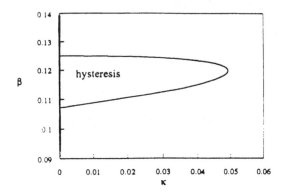

Figure 4.1. Hysteresis singularities - Model I ($\sigma = 1$, $\lambda = 1$).

The F_θ condition is evaluated numerically along with the hysteresis surface and only one point is found that it does not satisfy the $F_\theta \neq 0$ condition. The last constraint $F_{x\theta}$ is found to be non-zero for all kinds of singularities of this system for positive system parameters. This will be proven in the last part of this section.

(B)   *Isola and Mushroom*

The second possible qualitative change that can occur in the steady-state locus is the appearance of an isola and growth of an isola into a mushroom. The requirements for these two changes in the steady state behaviour have the form:

$$F = F_x = F_\theta = 0$$

with the additional requirements

$$F_{x\theta} \neq 0, \quad F_{xx} \neq 0, \quad F_{\theta\theta} \neq 0$$

For this model these conditions have the following forms:

$$F = \Sigma\Psi - x_{ss}(\Psi + \beta)^2 = 0 \qquad (4.14)$$

$$F_x = \Sigma\Psi_x - (\Psi + \beta)^2 - 2x_{ss}\Psi_x(\Psi + \beta) = 0 \qquad (4.15)$$

$$F_\theta = \Sigma_\theta\Psi = \Sigma\Psi_\theta - 2x_{ss}\Psi_\theta(\Psi + \beta) = 0 \qquad (4.16)$$

where

$$\Sigma_\theta = \frac{(\kappa\theta)^2 - 1}{\theta^2}$$

$$\Psi_\theta = (1 - x_{ss})\frac{\xi^2\lambda}{\theta^2}$$

First we combine the conditions $F = F_x = 0$ to express β in term of x_{ss} and other parameters by substituting the value of Σ from equation 4.14 into equation 4.15

$$\beta = \frac{-\Psi(\Psi + x\Psi_x)}{(\Psi - x\Psi_x)} \qquad (4.17)$$

Now we solve for the dimensionless decay rate constant κ. We process this by manipulating the conditions $F = F_x = \theta$ to express the quantity Σ_θ in term of x_{ss} and the other parameters:

$$\Sigma_\theta = \frac{-\Psi_\theta(\Psi + \beta)^2}{\Psi\Psi_x}$$

which when divided by Σ which is given by equation 4.14 gives:

$$\kappa = \frac{x\Psi_x - \theta\Psi_\theta}{\theta(x\Psi_x + \theta\Psi_\theta)} \qquad (4.18)$$

Next we substitute equation 4.17 and equation 4.18 for β and κ into the equation $F = 0$ to solve for x_{ss} in term of θ, σ, and λ. After a tedious manipulation the following quartic form for x_{ss} are obtained:

$$x_{ss}^4 + Ax_{ss}^3 + Bx_{ss}^2 + Cx_{ss} + D = 0 \qquad (4.19)$$

where

$$A = \frac{-(3\xi^2 - 4\sigma\xi + 2\xi - 3\sigma)\xi}{(\xi + 1)(\xi^2 - 2\sigma\xi - \sigma)}$$

$$B = \frac{-(-3\xi^2 + 8\sigma\xi^2 - \xi^2 + 4\sigma\xi - 5\sigma^2\xi - 3\sigma^2)\xi^2}{(\xi + 1)(\xi^2 - 2\sigma\xi - \sigma)^2}$$

$$C = \frac{-\xi^2}{(\xi + 1)(\xi^2 - 2\sigma\xi - \sigma)}$$

$$D = \frac{\sigma(-\xi + \sigma)\xi}{\lambda(\xi + 1)(\xi^2 - 2\sigma\xi - \sigma)^2}$$

Unlike equation 4.13 in the hysteresis case, a simple form for the isola-mushroom singularities in the κ, β-plane cannot be found. Therefore we need first to solve for x_{ss} from equation 4.19 then use the equation 4.17 and equation 4.18 to define the isola-mushroom boundaries in the κ, β-plane. For fixed σ and λ and for some θ values, two real solutions for x_{ss} of equation 4.19 are found; one of them corresponds to the birth of

57

an isola and the other to appearance of mushroom. This results in the appearance of two curves in the parameter space Figure 4.2. These singularities divide the parameter space into three regions: unique, isola and mushroom.

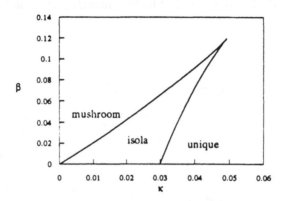

Figure 4.2. Isola singularities - Model I ($\sigma = 1$, $\lambda = 1$).

The combination of the hysteresis and the isola-mushroom varieties for fixed σ and λ in the κ, β-plane divides the parameter space into four regions of different multiplicity behaviour : unique, isola, mushroom and hysteresis. Figure 4.3 shows the complete branch diagram for singularities of codimension-one in the κ, β-parameter space for constant σ and λ.

The maximum value for β at which the system has multiple solutions can be found by observing that as the residence time approaches infinity, the quantity ξ approaches a constant value equal to σ. The maximum β can be found by substituting this value into equation 4.10, which gives

$$\beta_{max} = \frac{\sigma}{8}$$

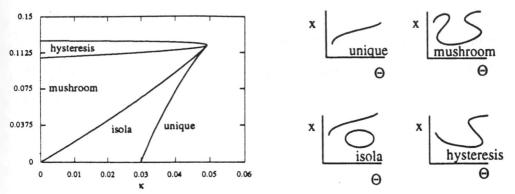

Figure 4.3. Branch set - Model I ($\sigma = 1$, $\lambda = 1$).

The maximum value for the decay rate constant κ at which the system has multiple solutions is the value at which the derivative of κ with respect to β of the hysteresis condition, eqation 4.13, is equal to zero. The effect of the parameters σ and λ on the locus of the turning point of the hysteresis curve $\left(d\kappa/d\beta = 0 \right)$ is shown in Figure 4.4.

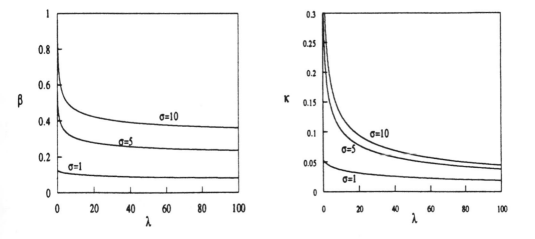

Figure 4.4. Effect of σ and λ on the turning point of the hysteresis curve.

The effect of the parameters σ and λ on the branch diagram is shown in Figures 4.5 and 4.6. From these figures we can see that increasing the parameter σ results in enlarging the multiplicity region, while increasing the parameter λ has an opposite effect. The effect of σ and λ can be studied by considering the limiting case when $\sigma = 0$ or $\lambda = \infty$. The mathematical model for this system is reduced to the following two-variable model:

$$0 = \frac{-x_{ss}}{\theta} - x_{ss}y_{ss}^2$$

$$0 = \frac{\beta - y_{ss}}{\theta} + x_{ss}y_{ss}^2 - \kappa y_{ss}$$

The steady state equation for this model is:

$$F(x_{ss}, \theta, \kappa, \beta) = \frac{(1 + \kappa\theta)^2 x_{ss}}{\theta} - x_{ss}(\beta - x_{ss})^2 = 0$$

The singularity condition is:

$$F_x = \frac{-(1 + \kappa\theta)^2}{\theta} - (\beta - x_{ss})^2 + 2x_{ss}(\beta - x_{ss})$$

Using these two conditions, we can show very easily that there are no singular points for this model in the positive κ, β-plane. In other words there is no multiplicity region in the κ, β-parameter space. Therefore when $\sigma = 0$ or $\lambda = \infty$ the first model has no multiplicity pattern for any the positive parameters. Another method is to consider the

60

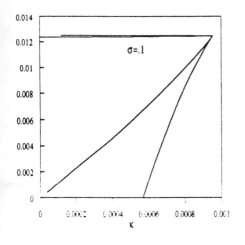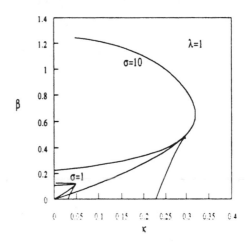

Figure 4.5. Effect of σ on the branch set.

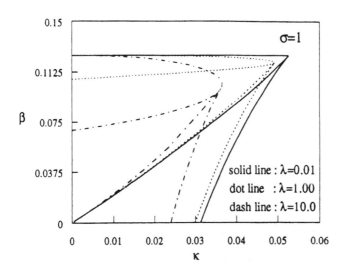

Figure 4.6. Effect of λ on the branch set.

effect of the parameters σ and λ in the quantity ξ. When $\sigma = 0$ or $\lambda = \infty$ the quantity ξ $\rightarrow 0$ and the hysteresis conditions become

$$x_{ss} = 0, \quad \kappa = \frac{-1}{\theta}, \quad \beta = 0$$

which shows that there is no multiplicity behaviour in our model for this case. Therefore we can conclude that decreasing σ or increasing λ reduces the multiplicity region in the parameter space.

For an infinite mass transfer coefficient $\lambda \rightarrow 0$, when the concentration of the component A in both tanks becomes the same, the hysteresis region is shrunk and the branch diagram is similar to the single CSTR case (Gray and Scott model) qualitatively. In this case the quantity ξ takes the form

$$\xi = \sigma$$

Therefore the parameter β at the hysteresis singularities has a constant value

$$\beta = \frac{\sigma}{8}$$

which is represented by a straight line in the κ, β-parameter plane. Only isola and mushroom behaviour can be obtained in this case.

The Gray and Scott (single CSTR) branch diagram can be found exactly by using the modified model presented in Section 2.2.1. We have seen that when $\sigma \rightarrow 0$, the branch diagram in $\kappa, \hat{\beta}$-parameter plane is equivalent to the branch diagram of the Gray and Scott model, see Figure 4.7.

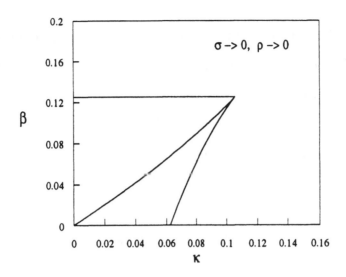

Figure 4.7. Gray and Scott limiting case.

4.2.2 Codimension-Two Singularities

Pitchfork bifurcation

The bifurcation diagram for the pitchfork singularity (Figure 3.4) shows two curves of equilibrium points intersected at the pitchfork singularity. Only one curve exists on both sides of the bifurcation point, however, its stability changes on passing through this point. The other curve lies entirely on one side of the pitchfork bifurcation point and has stability type that is opposite of the first curve on the other side. In order for the single scalar function to undergo a pitchfork bifurcation at (x_{ss}, θ), it is sufficient to have

$$F = F_x = F_\theta = F_{xx} = 0$$

and

$$F_{x\theta} \neq 0 \quad F_{xxx} \neq 0$$

Under these conditions, the function F is equivalent to a pitchfork bifurcation normal form and it provides a bifurcation diagram similar to Figure 3.4(0). If F is a universal unfolding of the pitchfork, then there are essentially four different bifurcation diagrams which can occur as the system parameters vary. These diagrams are illustrated in Figure 3.4. The pitchfork conditions for our model are :

$$F = \Sigma\Psi - x_{ss}(\Psi + \beta)^2 = 0 \tag{4.20}$$

$$F_x = \Sigma\Psi_x - (\Psi + \beta)^2 - 2x_{ss}\Psi_x(\Psi + \beta) = 0 \tag{4.21}$$

$$F_\theta = \Sigma_\theta\Psi + \Sigma\Psi_\theta - 2x_{ss}\Psi_\theta(\Psi + \beta) = 0 \tag{4.22}$$

$$F_{xx} = -4\Psi_x(\Psi + \beta) - 2x_{ss}\Psi_x^2 = 0 \tag{4.23}$$

(Σ, Ψ and their derivatives were defined above.)

In the hysteresis case, solving $F = F_x = F_{xx} = 0$ defines the reactant concentration x_{ss}, β and κ in terms of ξ:

$$x_{ss} = \frac{3}{4} \left. \frac{\xi}{\xi = 1} \right. \tag{4.24}$$

$$\beta = \frac{\xi}{8} \tag{4.25}$$

$$\Sigma = \frac{(1 + \kappa\theta)^2}{\theta} = \left(\frac{3}{4}\right)^3 \frac{\xi^2}{\xi + 1} \tag{4.26}$$

64

We substitute these expression into the fourth condition $F_\theta = 0$ to obtain another expression for κ in term of ξ:

$$\Sigma_\theta = \frac{(\kappa\theta)^2 - 1}{\theta^2} = \frac{9}{64}\frac{\lambda\xi^3(\xi + 4)}{(\xi + 1)^2} \qquad (4.27)$$

Dividing equation 4.26 by equation 4.27 and with some modifications, a simple expression for κ in term of ξ can be found:

$$\kappa = \frac{\big(3\sigma(\xi + 1) + (\sigma - \xi)(\xi + 4)\big)(\sigma - \xi)}{\big(3\sigma(\xi + 1) - (\sigma - \xi)(\xi + 4)\big)\sigma\lambda\xi} \qquad (4.28)$$

Substituting equation 4.28 into either equation 4.27 or equation 4.26 defines the pitchfork singularities in the θ, σ, λ-parameter space :

$$\lambda = \left(\frac{4}{3}\right)^3 \frac{36\sigma(\xi + 1)^3(\sigma - \xi)}{\xi^3\big(3\sigma(\xi + 1) - (\sigma - \xi)(\xi + 4)\big)^2} \qquad (4.29)$$

Note that $\lambda = 0$ when $\xi = \sigma$ (or $\beta = \sigma/8$) and $\lambda = \infty$ when the denominator of equation 29 equal to zero :

$$\xi = \frac{-(4 + 2\sigma) = \sqrt{(4 + 2\sigma)^2 + 4\sigma}}{2}$$

For example, when $\sigma = 1$ the values for β at which pitchfork singularities can be obtained lie in the range:

$$\frac{-3 + \sqrt{10}}{8} < \beta < \frac{1}{8}$$

The first nondegeneracy condition $F_{xxx} \neq 0$ is valid for all the pitchfork singularities since

$$F_{xxx} = -6\Psi^2 \neq 0$$

For the other condition, we will prove the following: $F_{x\theta} \neq 0$ for the pitchfork singularities for all the parameters which are physically acceptable. This condition has the following form:

$$F_{x\theta} = \Sigma_\theta \Psi_x + \Sigma \Psi_{x\theta} - 2\Psi_\theta(\Psi + \beta) - 2x_{ss}\left(\Psi_x \Psi_\theta + \Psi_{x\theta}(\Psi + \beta)\right)$$

We will rewrite this condition in terms of ξ using the following expressions:

$$\Psi = -(1 + \xi)x_{ss} + \xi = \frac{\xi}{4}$$

$$\Psi_x = -(\xi + 1)$$

$$\Psi_\theta = (1 - x_{ss})\xi_\theta = \frac{\xi^2 \lambda (\xi + 4)}{4\theta^2 (\xi + 1)}$$

$$\Psi_{x\theta} = -\xi_\theta = \frac{-\xi^2 \lambda}{\theta^2}$$

$$\Psi + \beta = \frac{\xi}{4} + \frac{\xi}{8} = \frac{3\xi}{8}$$

$$\Sigma = \frac{27}{64} \frac{\xi^2}{\xi + 1}$$

$$\Sigma_\theta = \frac{9}{64} \frac{\xi^3 \lambda (\xi + 4)}{\theta^2 (\xi + 1)^2}$$

and

$$x_{ss} = \frac{3\xi}{4(\xi + 1)}$$

Substituting these expressions into $F_{x\theta}$ yields

$$\left\{ \frac{-9}{64} \frac{\xi^3 \lambda (\xi + 4)}{\theta^2 (\xi + 1)^2}(\xi + 1) \right\} + \left\{ \frac{27}{64} \frac{\xi^2}{\xi + 1} \left(\frac{-\xi^2 \lambda}{\theta^2} \right) \right\} - 2 \left\{ \frac{\xi^2 \lambda (\xi + 4)}{4\theta^2 (\xi + 1)} \right\} - \left\{ 2 \frac{3\xi}{4(\xi + 1)} \right\}$$

$$\left\{ -(\xi + 1) \frac{-\xi^2 \lambda (\xi + 4)}{4\theta^2 (\xi + 1)} + \frac{-\xi^2 \lambda}{\theta^2} \left(\frac{3\xi}{8} \right) \right\} = 0$$

By taking out $\xi^3 \lambda / \theta^2 (\xi + 1)$ we see that $F_{x\theta}$ is equivalent to

$$-\frac{9}{64}(\xi + 4) - \frac{27}{64}\xi - \frac{3}{16}(\xi + 4) + \frac{3}{8}(\xi + 4) + \frac{9}{16}\xi = 0$$

The solution for $F_{x\theta}$ is given in term of ξ:

$$\xi = -1$$

But ξ is positive for positive θ, σ and λ, therefore the condition $F_{x\theta} \neq 0$ is valid for all the physically acceptable parameter values. We conclude that the nondegeneracy conditions for the pitchfork singularities for our system are satisfied.

Finally, we consider the recognition problem for the universal unfolding of the pitchfork singularities. For our problem the function F is a universal unfolding if the determinant of the following matrix is non-zero

$$\begin{pmatrix} 0 & 0 & F_{x\theta} & F_{xxx} \\ 0 & F_{x\theta} & F_{\theta\theta} & F_{\theta xx} \\ F_{\kappa} & F_{\kappa x} & F_{\kappa\theta} & F_{\kappa xx} \\ F_{\beta} & F_{\beta x} & F_{\beta\theta} & F_{\beta xx} \end{pmatrix}$$

We cannot prove that this matrix has full rank analytically because of the complexities of the derivatives of the function of F. Numerical methods are used to solve this problem. Figure 4.8 shows the pitchfork singularities in the β, λ-plane for different σ values. For example, all possible bifurcation diagrams for this system are obtained for the case when $\sigma = 10$ and $\lambda = 0.3025$. The branch set is shown in Figure 4.9. At the pitchfork point the bifurcation diagram is plotted in Figure 4.10a. Perturbing the system around this point leads to four possible multiplicity behaviours shown in Figure 4.10b, c, d, e. The unique solution can be found for system parameters which are close to these of the pitchfork point (Figure 4.10f).

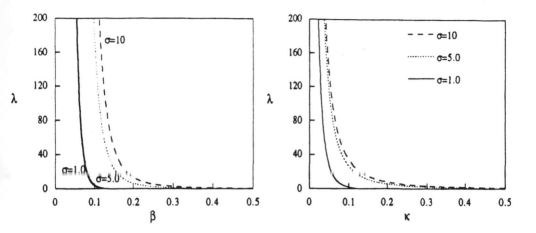

Figure 4.8. Pitchfork singularities in the λ, β- and λ, κ-spaces.

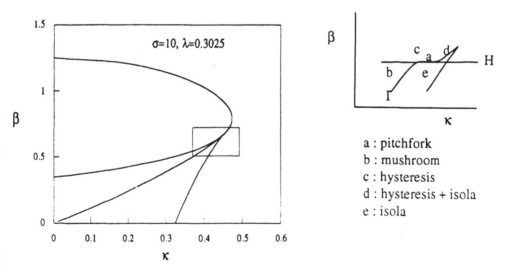

a : pitchfork
b : mushroom
c : hysteresis
d : hysteresis + isola
e : isola

Figure 4.9. Branch set ($\sigma = 10$, $\lambda = 0.3025$).

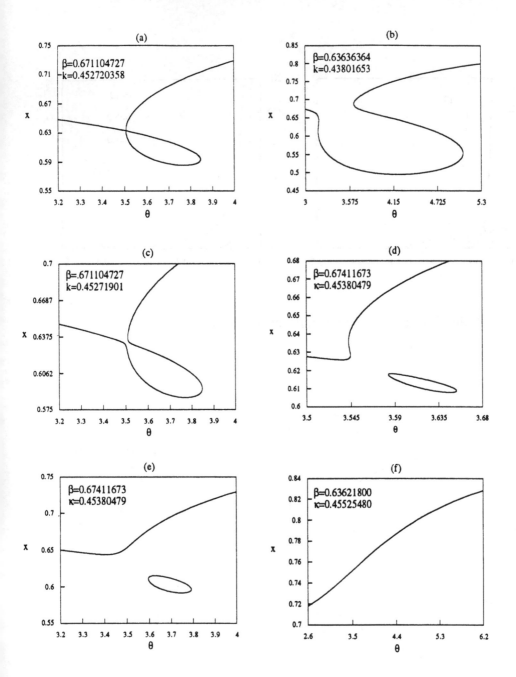

Figure 4.10. Bifurcation diagrams around the pitchfork singularity.

4.2.3   Codimension-three Singularities

Winged cusp bifurcation

The recognition problem for the winged cusp singularities is solved by the following conditions:

$$F = F_x = F_\theta = F_{xx} = F_{x\theta} = 0; \quad F_{xxx} > 0, F_{\theta\theta} > 0$$

The condition $F_{x\theta}$ is shown not equal to zero for positive system parameters. The nondegeneracy condition F_{xxx} is negative for all cases: $F_{xxx} = -6\Psi^2$. Therefore we conclude that this system does not have singularities equivalent to the winged cusp singularities.

Chapter 5

STEADY STATE ANALYSIS: MODEL II

The study of multiple solution of the steady state equations of the second model (coupling through the autocatalyst) will be reduced to studying how the solutions of a single scalar equation

$$F(x_{ss}, \theta) = 0$$

vary with the bifurcation parameter θ. In Section 5.2, we use the singularity theory to find all the multiplicity patterns of this case.

5.1   Steady State Relationship

The steady state model of this system is given as:

$$0 = \frac{1 - x_{ss}}{\theta} - x_{ss} y_{ss}^2 \tag{5.1}$$

$$0 = \frac{z_{ss} - y_{ss}}{\lambda} - \frac{y_{ss}}{\theta} + x_{ss} y_{ss}^2 - \kappa y_{ss} \tag{5.2}$$

$$0 = -\frac{z_{ss} - y_{ss}}{\lambda} + \frac{\sigma}{\theta} (\beta - z_{ss}) - \rho \kappa z_{ss} \tag{5.3}$$

The third equation gives a linear relationship between the concentration of the autocatalyst in the reactor, y_{ss} and its concentration in the tank, z_{ss}:

$$z_{ss} = \frac{y_{ss}\,\theta + \beta\sigma\theta}{\theta + \sigma\lambda + \rho\lambda\kappa\theta} \tag{5.4}$$

A linear relationship between x_{ss} and y_{ss} is found by substituting equation 5.4 into eqation 5.2 and adding equations (5.1) and (5.2):

$$y_{ss} = \frac{(1 - x_{ss})\,\phi + \beta\sigma\theta}{\sigma\theta + \rho\kappa\theta^2 + \phi(1 + \kappa\theta)} \tag{5.5}$$

where

$$\phi = \theta + \sigma\lambda + \rho\lambda\kappa\theta$$

Substituting equation 5.5 into equation 5.1 results in a single scalar equation:

$$F(x_{ss},\ \theta,\ \kappa,\ \beta,\ \sigma,\ \lambda,\ \rho) = \frac{1 - x_{ss}}{\theta} - x_{ss}\left(\frac{(1-x_{ss})\,\phi + \beta\sigma\theta}{\sigma\theta + \rho\kappa\theta^2 + \phi(1 + \kappa\theta)}\right)^{2} \tag{5.6}$$

The steady state solutions obtained from equation 5.6 depend on all the parameters. Equation 5.6 is a cubic in x_{ss}, so for a given value of θ, there will be either one or three solutions.   We will use singularity theory to define the hysteresis, isola-mushroom, pitchfork and winged cusp singularities in the parameter space.   The residence time θ is selected to be the bifurcation parameter.   The states of this system are bounded.   The reactant x_{ss} has a maximum value equal to one when there is no reaction.   The minimum value for x_{ss} is zero, which corresponds to complete conversion.   The maximum value for the autocatalyst can be found by setting $x_{ss} = 0$ and $\kappa = 0$ in the steady state model.

For this case equation 5.1 shows that θ tends to infinity when $x_{ss} = 0$. Under these conditions, equation 5.2 shows that the concentrations of the autocatalyst in both tanks are equal. Mass balance over the whole system yields:

$$y_{max} = \frac{1 + \sigma\beta}{1 + \sigma}$$

The maximum value of the state z_{ss} is either the same as y_{max} or the initial autocatalyst concentration β. To summarize, we must have the states of the system lying in the regions:

$$0 \leq x \leq 1, \quad 0 \leq y \leq \frac{1 + \beta\sigma}{1 + \sigma}, \quad 0 \leq z \leq max\left[\frac{1 + \beta\sigma}{1 + \sigma}, \beta\right]$$

5.2   Singularity Theory

5.2.1   Codimension-one Singularities

(A)   *Hysteresis*

The conditions for the appearance/disappearance of a hysteresis loop are given as:

$$F = F_x = F_{xx} = 0; \quad F_\theta \neq 0, \quad F_{x\theta} \neq 0, \quad F_{xxx} \neq 0$$

We rewrite the steady state equation equation 5.6 in the form:

$$F(x, \theta, \kappa, \beta, \sigma, \lambda,) = \Delta^2 (1 - x_{ss}) - x_{ss} \theta \Gamma^2 \tag{5.7}$$

where

$$\Delta = \sigma\theta + \rho\kappa\theta^2 + \phi\,(1 + \kappa\theta)$$

$$\Gamma = (1 - x)\,\phi + \beta\sigma\theta$$

We begin by computing the first and second derivative of F with respect to x:

$$F_x = -\Delta^2 - \theta\Gamma^2 - 2x_{ss}\,\theta\Gamma_\lambda\Gamma \tag{5.8}$$

$$F_{xx} = -4\Gamma - 2x_{ss}\Gamma_x \tag{5.9}$$

where

$$\Gamma_x = -\phi$$

Substituting the value of Δ^2 of equation 5.8 into the equation $F = 0$, we deduce

$$\Gamma = -2x_{ss}\,(1 - x_{ss})\,\Gamma_x \tag{5.10}$$

Substitution of equation 5.10 into $F_{xx} = 0$ (equation 5.9) yields

$$x_{ss} = \frac{3}{4} \tag{5.11}$$

Using this value of x_{ss} we manipulate the condition $F_{xx} = 0$ to express β in term of other parameters:

$$\beta = \frac{\phi}{8\sigma\theta} \tag{5.12}$$

Finally the decay rate constant can be found from the first condition by eliminating the x_{ss} and β defined by equation 5.11 and equation 5.12:

$$(\rho\lambda\theta^2)\kappa^2 + \left(\rho\theta^2 + \rho\lambda\theta + \sigma\lambda\theta + \theta^2 - \rho\lambda\theta\sqrt{27\theta/8}\right)$$

$$\kappa + \sigma\theta + (\theta + \sigma\lambda)(1 - \sqrt{27\theta/8}) = 0 \tag{5.13}$$

The equations for the hysteresis boundaries equation 5.11 - equation 5.13 are expressed in terms of the residence time θ and the other parameters (σ, λ, ρ). We should note that β is a function of κ so first we should solve equation 5.13 for κ, then use this value to define β fromequation 5.12. For $\sigma = 1$ and $\lambda = 1$, the hysteresis loci are shown in the κ, β-plane in Figure 5.1.

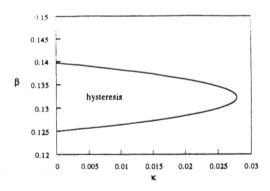

Figure 5.1. Hysteresis singularities - Model II ($\sigma = 1$, $\lambda = 1$, $\rho = 1$).

(B) Isola and Mushroom

The equations for the isola-mushroom boundaries are obtained from the conditions

$$F = F_x = F_\theta = 0, \quad F_{xx} \neq 0, F_{x\theta} \neq 0, F_{\theta\theta} \neq 0$$

For this model, these conditions take the form:

$$F = \Delta^2 (1 - x_{ss}) - x_{ss} \theta \Gamma^2 \tag{5.14}$$

$$F_x = -\Delta^2 - \theta \Gamma^2 - 2x_{ss} \theta \Gamma_x \Gamma \tag{5.15}$$

$$F_\theta = 2\Delta\Delta_\theta (1 - x_{ss}) - x_{ss} (\Gamma^2 + 2\theta\Gamma\Gamma_\theta) \tag{5.16}$$

where

$$\Delta_\theta = \sigma + 2\rho\kappa\theta + \phi_\theta (1 + \kappa\theta) + \phi\kappa$$

$$\Gamma_\theta = (1 - x_{ss}) \phi_\theta + \beta\sigma$$

$$\phi_\theta = 1 + \rho\lambda\kappa$$

Starting from the two conditions $F = F_x = 0$, we substitute the quantity Δ^2 from equation 5.14 into equation 5.15 to give:

$$\Gamma = -2x_{ss} (1 - x_{ss}) \Gamma_x$$

Using the values of Γ and Γ_x in term of the original model parameters we can express β in terms of x_{ss} and other parameters:

77

$$\beta = \frac{\phi(1-x)(2x-1)}{\sigma\theta} \tag{5.17}$$

Eliminating β from the definition of the quantity Γ results in:

$$\Gamma = 2x_{ss}(1-x_{ss})\phi,$$

then the first condition can be rewritten as:

$$\Delta^2 - 4\theta x_{ss}^3(1-x_{ss})\phi^2$$

Again rewriting this condition in terms of the original system parameters results in a quadratic equation for κ in term of x_{ss}, θ, σ, λ, and ρ:

$$(\rho\lambda\theta^2)\kappa^2 + \left(\rho\theta^2 + \rho\lambda\theta + \sigma\lambda\theta + \theta^2 - \rho\lambda\theta\sqrt{4\theta x_{ss}^3(1-x_{ss})}\right)\kappa$$

$$+ \sigma\theta + (\theta + \sigma\lambda)\left(1\sqrt{r\theta x_{ss}^3(1-x_{ss})}\right) = 0 \tag{5.18}$$

Substituting equaiton 5.17 and equation 5.18 for β and κ into the third condition (equation 5.16) results in a nonlinear equation which is sufficiently complicated so that to solve for x_{ss} or any other parameter in term of others is virtually impossible. A continuation algorithm is needed for this case to define the isola-mushroom boundaries in the parameter space. For $\sigma = 1$ and $\lambda = 1$ the isola-mushroom boundaries are shown in Figure 5.2.

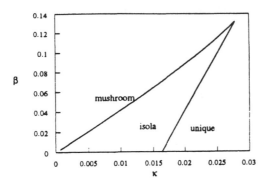

Figure 5.2. Isola singularities - Model II ($\sigma = 1$, $\lambda = 1$, $\rho = 1$).

The combination of the hysteresis and the isola-mushroom varieties for $\sigma = 1$ and $\lambda = 1$ is shown in Figure 5.3. The $\kappa - \beta$ plane is divided into four regions: unique, isola, mushroom and hysteresis patterns.

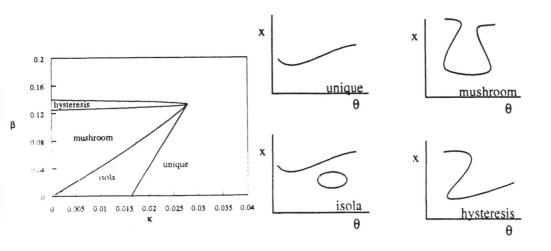

Figure 5.3. Branch set - Model II ($\sigma = 1$, $\lambda = 1$, $\rho = 1$).

The maximum value for β at which multiplicity occurs, is obtained by setting $\kappa = 0$ in equation 5.13:

$$\sigma\theta + (\theta + \sigma\lambda)\left(1 - \sqrt{27\theta}/8\right) = 0$$

Using equation 5.12, we can write this relation in term of β:

$$f(\beta,\sigma,\lambda) = \beta^3 + \left(\frac{128\sigma - 27\sigma\lambda - 64}{512\sigma}\right)\beta^3 + \left(\frac{8\sigma - 16}{512\sigma}\right)\beta - \frac{1}{215\sigma} = 0 \qquad (5.19)$$

First we see that the parameter ρ does not have any affect on the locus of the intersections of the hysteresis curve with the line $\kappa = 0$. Equation 5.19 might have either one or three solutions. Solving the conditions $(f = f_\beta = f_{\beta\beta} = 0)$ gives us the point where the number of intersections changes from one to three. The solution for these conditions is:

$$\beta = 1/16, \quad \sigma = 8, \quad \lambda = 8$$

At this point the value of β_{max} changes in a dramatic way. For example for $\sigma = 8$, $\lambda = 7$, β_{max} equals 0.03 while for $\sigma = 8$, $\lambda = 9$ the value of β_{max} is 0.18. The loci of the solutions of equation 5.19 are shown in Figure 5.4. The hysteresis boundaries for the cases $(\rho = 1, \sigma = 7, 8, 9)$ are shown in Figure 5.5. From this we conclude that increasing λ, or decreasing σ, increases the value of β_{max} at which the system has multiple solutions.

The maximum value for κ where the system has more than one solution is found to be at the turning point of the hysteresis curve $(d\kappa,d\beta = 0)$. For the case $\rho = 1$, Figure 5.6 shows the effect of using different σ's or λ's on the location of the turning

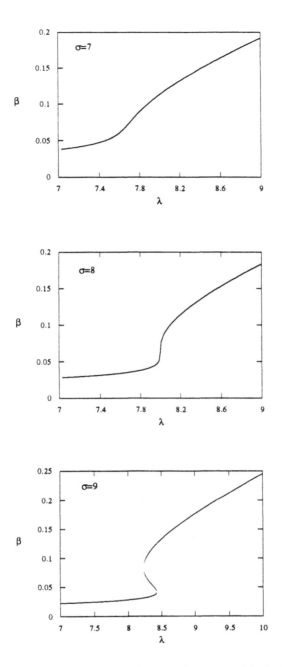

Figure 5.4. The loci of intersection of the hysteresis curve with the lines $\kappa = 0$ - Model II.

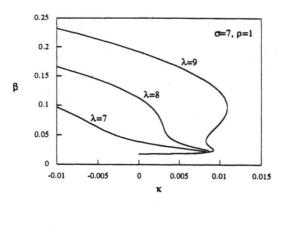

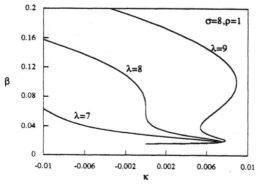

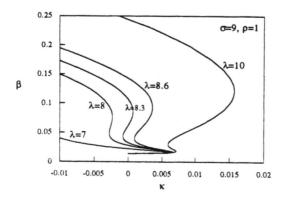

Figure 5.5. The hysteresis boundaries - Model II ($\rho = 1$).

82

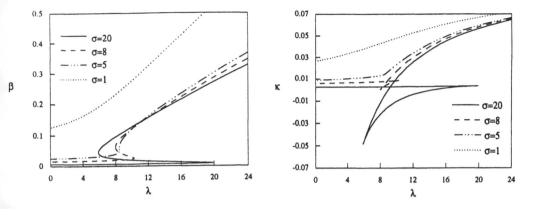

Figure 5.6. The effect of σ and λ on the loci of the turning points of the hysteresis curve - Model II ($\rho = 1$).

point. When we increase λ or decrease σ, the values of κ_{max} and $\beta_{turning\ point}$ increase. We have found that three turning points exist for some intermediate values for the parameters σ and λ. Figure 5.7 shows the effect of changing the parameter ρ on the loci of the turning points when $\sigma = 9$. At low λ values, increasing ρ reduces the values of κ_{max} at which multiple solution can be found, while for large λ increasing ρ increases the value of β at the turning point but it has a little effect on the value of κ_{max}. The appearance of more than one turning point in the branch diagram gives different looking hysteresis curves. Figure 5.8 shows qualitatively how these different curves arise. Starting from the case ($\sigma = 1, \lambda = 1, \rho = 1$), the following three figures show the effect of changing these parameters one at a time. Figures 5.9a, b and 5.10 show that the parameters σ and λ have opposite effects to those of the first model which was examined in Chapter 4. When σ increases the multiplicity region is reduced while the opposite happens when λ increases. Figure 5.11 shows that increasing the volume ratio parameter

ρ reduces the multiplicity region with respect to κ but it does not change the coordinate of β.

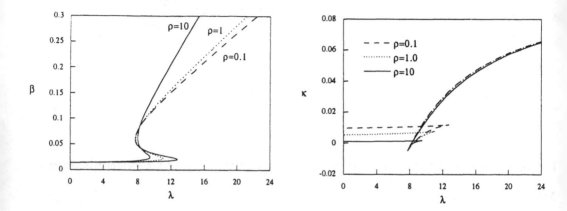

Figure 5.7. The effect of ρ and λ on the loci of the turning points of the hysteresis curve - Model II ($\sigma = 9$).

For an infinite mass transfer coefficient ($\lambda \rightarrow 0$) the concentrations of the autocatalyst in the tank and the reactor become the same and the branch diagram obtained for this case is equivalent to the single CSTR, where the hysteresis region is represented by a straight line in the κ, β–plane. From equation 5.12, this straight line is defined by

$$\beta = \frac{1}{8\sigma}$$

Note that the quantity ϕ approaches θ as the mass transfer resistance λ goes to zero.

This model can be also reduced to the single CSTR case by considering the limiting case $\sigma \rightarrow 0$ and $\rho \rightarrow 0$ in the modified model given in Section 2.3.1. The branch diagram for this case is shown in Figure 5.12.

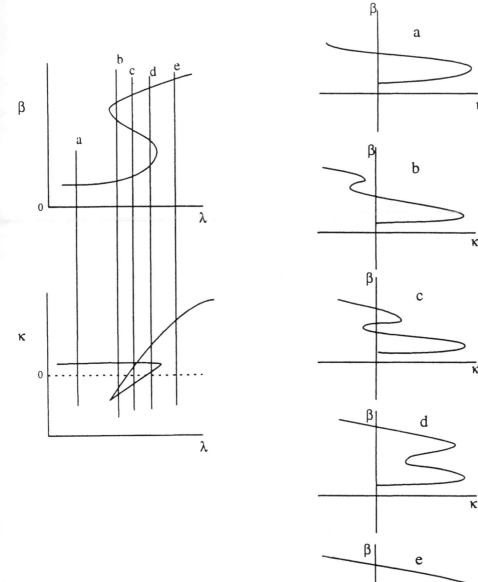

Figure 5.8. Hysteresis curves - Model II.

85

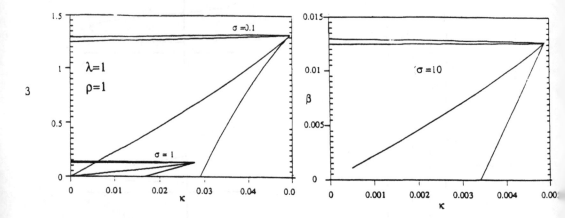

Figure 5.9. Effect of σ on the branch set - Model II.

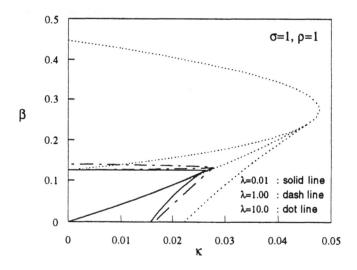

Figure 5.10. Effect of λ on the branch set - Model II.

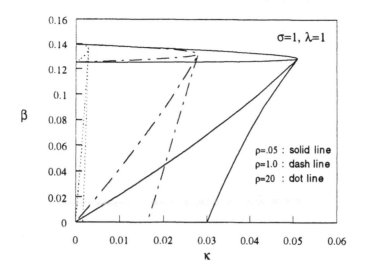

Figure 5.11. Effect of ρ on the branch set - Model II.

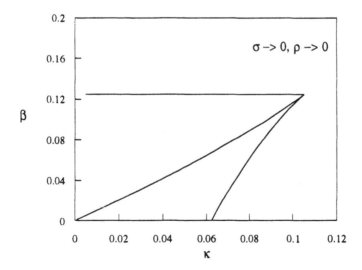

Figure 5.12. Branch set for Gray and Scott case.

5.2.2 Codimension-two Singularities

Pitchfork Bifurcation

The pitchfork conditions for the second system are the vanishing of:

$$F = \Delta^2 (1 - x_{ss}) - x_{ss} \, \theta \Gamma^2 \tag{5.20}$$

$$F_x = -\Delta^2 - \theta \Gamma^2 - 2 x_{ss} \, \theta \Gamma_x \Gamma \tag{5.21}$$

$$F_{xx} = -4\Gamma - 2 x_{ss} \, \Gamma_x \tag{5.22}$$

$$F_\theta = 2(1 - x_{ss}) \, \Delta \Delta_\theta - x_{ss} \Gamma^2 - 2 x_{ss} \, \theta \Gamma \Gamma_\theta \tag{5.23}$$

where Δ, Γ and their derivatives were defined before. We solve for x_{ss}, β, κ and λ in term of the parameters θ, σ, ρ. The solutions for the hysteresis problem given in the last section satisfy the first three conditions of the pitchfork singularity. They are

$$x_{ss} = 3/4$$

$$\beta = \frac{\phi}{8\sigma\theta}$$

$$\rho\lambda\theta^2\kappa^2 + \left(\rho\theta^2 + \rho\lambda\theta + \sigma\lambda\theta + \theta^2 - \rho\lambda\theta\sqrt{27\theta}/8\right)\kappa$$

$$+ \left(\sigma\theta + (\theta + \sigma\lambda)(1 - \sqrt{27\theta})/8\right) = 0$$

Using the hysteresis expressions we use the last condition F_θ (equation 5.23) to define the pitchfork boundaries in the θ-λ-σ-ρ parameter space:

$$\lambda = \frac{16\sqrt{27\theta}\,(\sigma + 2\rho\kappa\theta + 1 + 2\kappa\theta) - 81\,\theta}{81\rho\kappa\theta + 45\,\sigma - \sqrt{27\theta}\,(\sigma\kappa + \rho\kappa + 2\rho\kappa^2\theta)/8} \tag{5.24}$$

88

where κ is the solution for the quadratic equation (equation 5.13). Since κ also is function of λ, equation 5.24 and the quadratic equation 5.13 need to be solved simultaneously. For fixed σ and ρ the pitchfork singularities can be plotted as λ against β by substituting the value of θ from Eequation 5.12 into equqtion 5.24. Figure 5.13 shows the pitchfork singularities in the λ-β plane. The corresponding values x_{ss}, θ and κ can be found from the expressions equation 5.11, equation 5.12, and equation 5.13. Figure 5.13 shows that for some σ's and ρ's, multiple pitchfork points exist in the branch diagram. Plotting the turning point locations in Figure 5.13 against σ for different ρ divides the parameter space into two regions shown in Figure 5.14. In the first region one pitchfork appears in the branch diagram, while in the second part three pitchfork points appear in the branch diagram. For $\rho = 1$ the branch diagram has one pitchfork point if the system is operated with $\sigma < 11.40127$, and it has three pitchfork points when σ is bigger than 11.40127. For $\rho = 10$ the critical value corresponds to $\sigma = 8.535032$. For example, when $\sigma = 17.79479$ and $\rho = 1$, at $\lambda = 25$, three points satisfy the pitchfork conditions, namely:

$$x = 0.75, \quad \beta = 0.058592560, \quad \kappa = 0.02912163, \quad \theta = 67.27092$$
$$x = 0.75, \quad \beta = 0.015611298, \quad \kappa = 0.00712135, \quad \theta = 425.9716$$
$$x = 0.75, \quad \beta = 0.011485870, \quad \kappa = 0.00520934, \quad \theta = 881.1465$$

The bifurcation diagrams for these cases are shown in Figure 5.15.

Now we examine the nondegeneracy conditions. The first condition $F_{xxx} \neq 0$ is valid for all positive parameters since

$$F_{xxx} = -6\Gamma_x = 6\phi, \quad \phi > 0$$

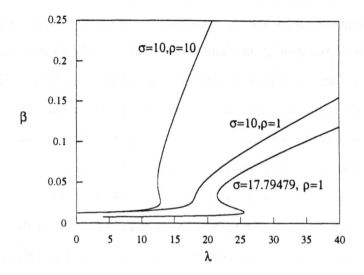

Figure 5.13. Pitchfork singularities - Model II.

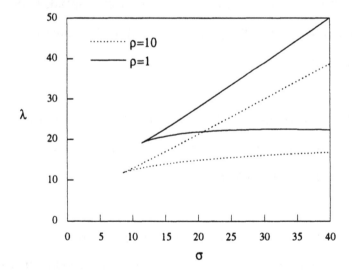

Figure 5.14. Loci of the turning points of the pitchfork singularities - Model II.

90

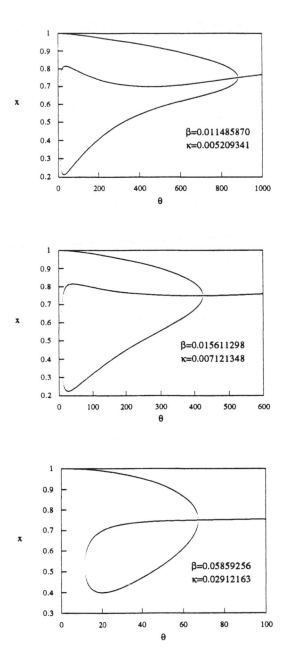

Figure 5.15. Bifurcation diagrams for three pitchfork singularities (σ = 17.79479, λ = 25.0, ρ = 1.0 - Model II.

91

Now we will show that the condition $F_{xxx} \neq 0$ is valid for all the parameters which are physically acceptable. $F_{x\theta}$ takes the form:

$$F_{x\theta} = -2\Delta\Delta_\theta - \Gamma^2 - 2\theta\Gamma\Gamma_\theta - 2x_{ss}\Gamma_x\Gamma - 2x_{ss}\theta(\Gamma_\theta\Gamma_x + \Gamma\Gamma_{x\theta})$$

First we use the equality $F_{x\theta} = 0$ to express the quantity $\Delta\Delta_\theta$ in terms of x, θ, Γ and Γ derivatives. Then we substitute this in the condition $F_\theta = 0$ to get:

$$F_\theta = -\Gamma^2 - 2\theta\Gamma\Gamma_\theta - (1 - x_{ss})\left(2x_{ss}\Gamma\Gamma_x + 2x_{ss}\theta(\Gamma_\theta\Gamma_x + \Gamma\Gamma_{x\theta})\right) = 0$$

The conditions $F = F_x = F_{xx} = 0$ give the following expressions:

$$x_{ss} = 0.75$$

$$\Gamma_x = -\frac{8}{3}\Gamma, \quad \left(\beta = \frac{\phi}{8\sigma\theta}\right)$$

These expressions are used to replace x_{ss} and Γ_x in $F_\theta = 0$. By rearranging terms, $F_\theta = 0$ can be reduced to the following simple equation:

$$F_\theta = -\Gamma_\theta - \frac{3}{8}\Gamma_{x\theta} = 0$$

where the derivatives of Γ are

$$\Gamma_\theta = (1 - x_{ss})\,\phi_\theta + \beta\sigma$$
$$\Gamma_{x\theta} = \phi_\theta$$

Therefore the condition $F_{x\theta} = 0$ leads to the expression:

$$\beta = \phi_\theta/8$$

But we recall that the pitchfork points have the condition

$$\beta = \phi/8\theta\sigma$$

which satisfies the $F_{x\theta} = 0$ condition if

$$\phi = \theta\phi_\theta$$

or

$$\sigma\lambda = 0$$

This proves that the condition $F_{x\theta} \neq 0$ is valid for all the singularities defined in the positive parameter space.

Finally we should check the determinant of the universal unfolding matrix for pitchfork singularities. If the the matrix has full rank, then perturbing the system around the pitchfork point leads to four bifurcation diagrams. Consider the following numerical example: let $\sigma = 1$, $\lambda = 10$ and $\rho = 1$. By perturbing the system around the pitchfork point ($\beta = 0.2393087$, $\kappa = 0.04550854$) four bifurcation diagrams result (Figure 5.16b, c, d, e). Close to this pitchfork point the unique solution (Figure 5.16f) is found.

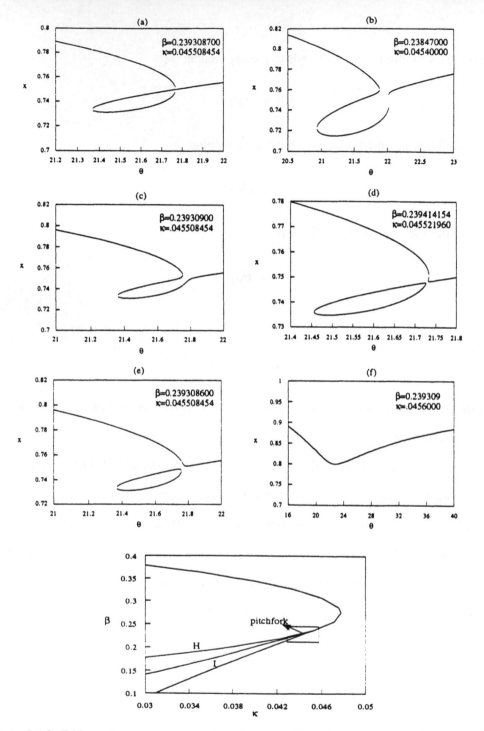

Figure 5.16. Bifurcation diagrams obtained by perturbing the model around the pitchfork point ($\sigma = 1.0$, $\lambda = 10.0$, $\rho = 1.0$) - Model II.

5.2.3   Codimension-three Singularities

Winged Cusp Bifurcation

The conditions for the winged cusp singularities are:

$$F = F_x = F_\theta = F_{xx} = F_{x\theta} = 0, \quad F_{xxx} > 0, F_{\theta\theta} > 0$$

We have shown that $F_{x\theta} = 0$ is not valid for positive parameters. This shows that the second model does not have singularities equivalent to the winged cusp singularities.

Chapter 6

DYNAMICS - THEORY

In this chapter we study the periodic solutions of our nonlinear systems, for, in contrast to linear systems, it is possible to have structurally stable periodic oscillations. Hopf [1983] proved the existence of periodic solutions associated with a change in the stability of singular points that occurs when a single pair of eigenvalues of the linearized n-dimensional system crosses the imaginary axis. This chapter is organized as follows. The classical Hopf bifurcation theory is introduced briefly in the first section. In the second section, we consider the interactions of static and dynamic bifurcations. In the third section we concentrate on the degeneracies when the second and the third hypotheses of Hopf theory break down. We present the results of the singularity theory developed by Golubitsky and Langford [1981] and Golubitsky and Schaeffer [1985] to study the dynamics near these degeneracies. Finally we use the Poincaré map to define the conditions for the stability and the secondary bifurcations of the periodic solutions.

6.1   Hopf Bifurcation Theorem

Consider the parametrized system:

$$\frac{dx}{dt} = F(x, \theta) \tag{6.1}$$

$$x \in IR^n, \ \theta \in IR^1$$

F is a C^r function on $IR^n \ x \ IR^1$, x is the state-variable vector and θ is the bifurcation parameter. The first hypothesis if the Hopf Theorem is:

H1: At (x_o, θ_o), the Jacobian matrix $DF_x(x,\theta)$ has simple eigenvalues $\pm i\omega$ and $DF_x(x,\theta)$ has no other eigenvalues on the imaginary axis.

Assume that the point (x_o, θ_o) has been transformed to the origin $(0, \theta_o)$. We will follow the analysis given by Wiggins [Wiggin, 1990] who used center manifold theory to show that the orbit structure near the nonhyperbolic point $(0, \theta_o)$ is determined by the following two-dimensional center manifold:

$$\begin{pmatrix} \dot{x}_1 \\ \dot{x}_2 \end{pmatrix} = \begin{pmatrix} \textbf{\textit{Re}}\ \mu(\theta) & \textbf{\textit{-Im}}\ \mu(\theta) \\ \textbf{\textit{Im}}\ \mu(\theta) & \textbf{\textit{Re}}\ \mu(\theta) \end{pmatrix} \begin{pmatrix} x_1 \\ x_2 \end{pmatrix} + \begin{pmatrix} f^1(x_1,x_2,\theta) \\ f^2(x_1,x_2,\theta) \end{pmatrix} \tag{6.2}$$

where f^1 and f^2 are nonlinear functions in x_1 and x_2, μ and $\bar{\mu}$ are the eigenvalues of equation 6.1 at the point (x_o, θ_o):

$$\mu(\theta) = v(\theta) + i\omega(\theta)$$

The first Hopf theorem hypothesis means:

$$v(\theta_o) = 0, \quad \omega(\theta_o) \neq 0$$

The next step is to use the normal form method to define the simplest mathematical form for equation 6.2. This method is based on the idea of introducing successive coordinate transformations to simplify the analytic expression of the vector field [Nayfeh, 1993;

Guckenheimer and Holmes, 1983; Wiggins, 1990]. A normal form for equation 6.2 has been found[Guckenheimer and Holmes, 1983; Wiggins, 1990]:

$$\begin{pmatrix} \dot{x}_1 \\ \dot{x}_2 \end{pmatrix} = \begin{pmatrix} v(\theta) & -\omega(\theta) \\ \omega(\theta) & v(\theta) \end{pmatrix} \begin{pmatrix} x_1 \\ x_2 \end{pmatrix} + \begin{pmatrix} (a(\theta)x_1 + b(\theta)x_2)(x_1^2 + x_2^2) + H.O.T. \\ (b(\theta)x_1 + a(\theta)x_2)(x_1^2 + x_2^2) + H.O.T. \end{pmatrix} \tag{6.3}$$

where a and b are constants depending on θ. In polar coordinates, which are more convenient, the normal form is:

$$\dot{r} = v(\theta)r + a(\theta)r^3 + O(r^5)$$
$$\dot{\theta} = \omega(\theta) + b(\theta)r^2 + O(r^4) \tag{6.4}$$

Since we are interested in the dynamics near $\theta = \theta_o$, it is natural to expand the coefficients v and ω in equation about $\theta = \theta_o$:

$$\dot{r} = v(\theta_o)\,\theta r + a(\theta_o)r^3 + O(\theta_o^2 r,\,\theta r^3,\,r^5)$$
$$\dot{\theta} = \omega(\theta_o) + \omega'(\theta_o) + b(\theta_o)r^2 + O(\theta^2,\,\theta r^2,\,r^4) \tag{6.5}$$

where prime (') denotes differentiation with respect to θ. Neglecting the higher order terms in equation 6.5 gives:

$$\dot{r} = v'(\theta_o)\,\theta r + a(\theta_o)r^3$$
$$\dot{\theta} = \omega(\theta_o) + \omega'(\theta_o) + b(\theta_o)r^3 \tag{6.6}$$

Values of $r > 0$ and θ for which $\dot{r} = 0$, but $\dot{\theta} \neq 0$, correspond to periodic solutions of equation 6.1 given in the following lemma:

Lemma *For* $-\infty < \dfrac{\theta v'}{a} < 0$ *and* θ *sufficiently small*

$$\left(r(t), \theta(t) \right) = \left(\sqrt{\frac{- \theta v'(\theta_0)}{a(\theta_0)}}, \left(\omega(\theta_0) + \left(\omega'(\theta_0) - \frac{b(\theta_0)\, v'(\theta_0)}{a(\theta_0)} \right) \theta \right) t + \theta_0 \right)$$

is a periodic orbit for equation 6.6.

The stability of these orbits can be examined by checking the sign of the coefficient a [72].

Lemma *The periodic orbit is asymptotically stable for negative a and unstable for positive a.*

The case $a < 0$ is referred to as a supercritical bifurcation where stable limit cycles emerge at the Hopf point. The second case $a < 0$ is known as a subcritical bifurcation where unstable periodic solutions are formed. The direction of bifurcation is determined by the signs of a and v'. Figure 6.1 shows the stability and the direction of periodic solutions in the phase plane.

The second hypothesis of Hopf theorem or transversality is:

$$v'(\theta_0) \neq 0$$

This condition guarantees that the real part of the complex eigenvalues changes its sign when passing through the Hopf bifurcation point $(0, \theta_0)$.

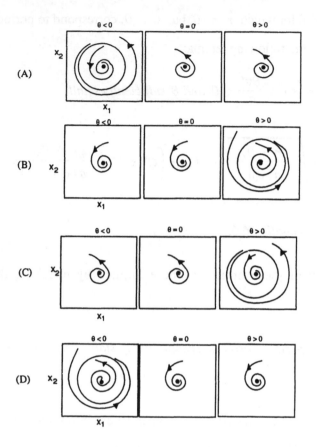

Figure 6.1. Periodic Solutions: A) $v' > 0$, $a > 0$; B) $v' > 0$, $a < 0$; C) $v' < 0$, $a > 0$; D) $v' < 0$, $a < 0$

The third hypothesis which needs to be satisfied in order for a periodic solutions to exist is called the stability hypothesis:

H3:

$$a(\theta_o) \neq 0$$

This condition describes the boundary at which the limit cycles change their stability.

100

Hopf Theorem

Consider the system given by equation 6.1 with

$$F(0, \theta_o) = 0$$

If **H1**, **H2**, and **H3** hold, then there exists a unique branch of periodic solutions emanating from $(0, \theta_o)$.

Evaluating the first and the second hypothesis is relatively straightforward, but calculating the stability coefficient involves more intricate algebraic manipulations. The explicit calculations are given in [Marsden and McCracken, 1976; Guckenheimer and Holmes, 1983], here we just state the result. At the Hopf bifurcation, equation 6.2 becomes

$$\begin{pmatrix} \dot{x_1} \\ \dot{x_2} \end{pmatrix} = \begin{pmatrix} 0 & -\omega \\ \omega & 0 \end{pmatrix} \begin{pmatrix} x_1 \\ x_2 \end{pmatrix} + \begin{pmatrix} f^1(x_1, x_2, \theta_o) \\ f^2(x_1, x_2, \theta_o) \end{pmatrix} \tag{6.7}$$

and the stability coefficient $a(\theta_o)$ is given by:

$$a = \frac{1}{16}\left(f^1_{111} + f^1_{122} + f^2_{112} + f^2_{222}\right) +$$

$$\frac{1}{16\omega}\left(f^1_{12}\left(f^1_{11} + f^2_{22}\right) - f^2_{12}\left(f^2_{11} + f^2_{22}\right) - f^1_{11}f^2_{11} + f^1_{22}f^2_{22}\right)$$

where $f_{111} = f_{x_1 x_1 x_1}$ and so on. To use this formula we should keep track of the coefficients in the normal form transformation in terms of our original system parameters. The details of the normal form method and the center manifold theory can be found in

Guckenheimer and Holmes [1983] or Wiggins [1990]. A general implicit formula for the stability coefficient has been derived by Golubitsky and Langford [1981]. Their result will be presented in the third section. It is found that the Hopf bifurcation is generic in one-parameter problems, i.e., it persists if the underlying model undergoes a small change. This means that Hopf bifurcation is of codimension zero.

6.2   Type-one Degenerate Bifurcations

This section examines the case where the first hypothesis of Hopf theorem fails. This occurs when more than the simple or complex conjugate pair of eigenvalues of the Jacobian simultaneously has zero real part. Consider the dynamical system of the form

$$\dot{x} = F(x, \theta, \alpha), \quad x \in IR^n, \theta \in IR, \alpha \in IR^m \tag{6.8}$$

where x is a state variable vector, θ is the bifurcation parameter and α is a constant vector of parameters. The steady state solutions of equation 6.8, x_{ss}, is found by setting the left side of equation 6.8 equal to zero. If we let $y = x - x_{ss}$, equation 6.8 is brought to a local form

$$\dot{y} = D_x F(x_{ss}, \theta, \alpha) + G(y, \theta, \alpha), \quad y \in IR^n$$

G contain terms at least quadratic in y. Suppose that at $(x_{ss,o}, \theta_o, \alpha_o)$ the total number of eigenvalues of the Jacobian with real part zero is N, then we may apply the center manifold theory and the normal form method to obtain a reduced N dimensional system which contains the local dynamic features of the original dynamic system:

$$\dot{v} = Jv + \Phi(v, \theta, \alpha)$$

The matrix J is of size $N \times N$. It is in Jordan form and its eigenvalues have zero real part at $(v_{ss}, \theta_o, \alpha_o)$. The method of normal forms is used to explain why a model of high dimensionality has an observable behaviour similar to that of low-dimensional models [Guckenheimer and Holmes, 1983; Lyberatos, Kuzsta, and Baily, 1985; Wiggins, 1990].

The different types of bifurcation (static, Hopf, and the first few members of the first type of degeneracies) are shown in Table 6.1 in terms of the Jordan block of the matrix J. The equivalent conditions for these bifurcations are given in terms of the coefficients of the characteristics equation of the Jacobian form. A three-dimensional system is considered in Table 6.1, and the codimensions for these degenerate bifurcations are given. These conditions may be used to define the parameter values for which these singularities can be found. Figure 6.2 shows how the different cases in Table 6.1 are obtained [Troger and Steindl, 1991]. The numbers 1 to 5 in each plot denotes five successive positions of the motion of the eigenvalues of interest in the complex plane under the the variation of the bifurcation parameter.

Any model that has eigenvalues of the steady-state Jacobian on the imaginary axis is structurally unstable. A small change in one or more parameter values changes the eigenvalue configuration and the topology of the local dynamics. When static bifurcation occurs, it involves steady-state bifurcation and leads to multiple steady states. At Hopf bifurcation the steady state becomes unstable, and a periodic branch, which may be stable or unstable, appears. $\mathbf{F_1}$ degeneracies result from the interactions of Hopf point and a static bifurcation point. Here the imaginary part of the complex-conjugate eigenvalue pair associated with the Hopf point goes to zero as the Hopf point and the

Bifurcation type	Jordan block structure	Algebraic conditions $-\mu_3 + S_1\mu^2 - S_2\mu + S_3 = 0$	Codimension
Static	(0)	$S_3 = 0$	0
Hopf	$\begin{pmatrix} 0 & -\omega \\ \omega & 0 \end{pmatrix}$	$S_1 S_2 - S_3 = 0,\ S_2 > 0$	0
Double-zero F_1	$\begin{pmatrix} 0 & 1 \\ 0 & 0 \end{pmatrix}$	$S_2 = S_3 = 0$	1
Pure imaginary eigenvalues and a simple zero F_2	$\begin{pmatrix} 0 & -\omega & 0 \\ \omega & 0 & 0 \\ 0 & 0 & 0 \end{pmatrix}$	$S_1 = S_3 = 0,\ S_2 > 0$	1
Triple-zero G_1	$\begin{pmatrix} 0 & 1 & 0 \\ 0 & 0 & 1 \\ 0 & 0 & 0 \end{pmatrix}$	$S_1 = S_2 = S_3 = 0$	2

Table 6.1. Degenerate bifurcation - type-1

limit point (saddle-node) come together. Close to F_1 degeneracies, one can expect steady-state multiplicity and periodic solutions. One might also observe some global bifurcation phenomena, such as a homoclinic orbit which connects a steady state to itself by a closed orbit in the phase space[Guckenheimer and Holmes, 1983; Wiggins 1990].

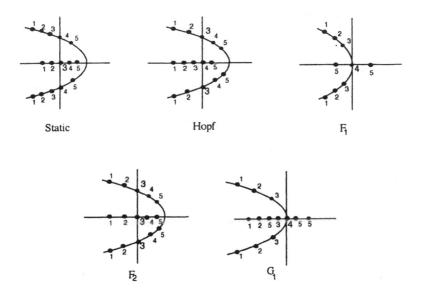

Figure 6.2. Eigenvalue movements in the complex plane

A pure imaginary eigenvalue and a simple zero F_2 is obtained from the Hopf and static bifurcations interactions. In this case the Hopf point remains a Hopf point as it passes through the limiting point so that three eigenvalues are at criticality. In addition to static bifurcation, periodic solutions and homoclinic orbits, one can anticipate that, close to an F_2 point, heteroclinc orbits (a closed orbit connects distinct fixed points), and toroidal oscillations may occur [Langford, 1979; 1973; Keener, 1979; Wiggins 1990]. The triple-zero eigenvalue case G_1 may produce anything ranging from chaos to steady-state multiplicity. The analysis of G_1 bifurcation is not complete [Lyberatos, Kuzsta, and Baily, 1985; Pismen, 1985].

6.3   Type-two and Type-three Degeneracies

In this section we study the degenerate Hopf bifurcations that arise when the second and the third hypotheses are not satisfied. The first family of type-2 degeneracies arises when the complex eigenvalues encounter the imaginary axis tangentially, or when $v' = 0$, while the first member of the third type occurs when the stability coefficient vanishes. We introduce the following notations in seeking a general definition for these types of degeneracies. The Taylor expansion of the normal form of any model at Hopf point is given:

$$\dot{r} = v'(\theta_o)\,\theta r + a_1(\theta_o)\,r^3 + a_2(\theta_o)\,r^5 + \ldots$$
$$\dot{\theta} = \omega(\theta_o) + \ldots \tag{6.9}$$

where $a_1 = a$, the stability coefficient given in the first section. Suppose that the bifurcation and the auxiliary parameters are arranged such that at the Hopf point we have:

$$v = v' = \ldots = v^k = 0, \quad v^{k+1} \neq 0 \tag{6.10}$$
$$a_1 = a_2 = \ldots = a_m = 0, \quad a_{m+1} \neq 0 \tag{6.11}$$

We use the notation $\mathbf{H_{km}}$ for the degenerate Hopf points which satisfy the conditions equations 6.10 and 6.11. The codimension of these degeneracies is $k + m + 1$. Table 6.2 summarizes the conditions for the first two families of type-2 and type-3 degeneracies.

Singularity theory [Golubitsky and Langford, 1981] is used to study the dynamics around the Hopf degenerate bifurcations $\mathbf{H_{01}}$, $\mathbf{H_{02}}$, $\mathbf{H_{10}}$, $\mathbf{H_{20}}$, $\mathbf{H_{11}}$. It has been shown that these degeneracies could be analyzed using the Lyapunov-Schmidt reduction method

Bifurcation Type	Conditions	Codimension
H_{00}	$v = 0, v' \neq 0, a_1 \neq 0$	0
H_{01}	$v = v' = 0, v" \neq 0, a_1 \neq 0$	1
H_{02}	$v = v' = v" = 0, v''' \neq 0, a_1 \neq 0$	2
H_{10}	$v = 0, a_1 = 0, v' \neq 0, a_2 \neq 0$	1
H_{20}	$v = 0, a_1 = a_2 = 0, v' \neq 0, a_3 \neq 0$	2
H_{11}	$v = v' = 0, a_1 = 0, v" \neq 0, a_2 \neq 0$	2

Table 6.2. Degenerate bifurcation - type-2 and type-3

and symmetry to reduce the bifurcation problem to a single implicitly defined equation. Then, similarly to the analysis of the steady-state, singularity theory can be used to analyze the solutions of this equation near the degenerate Hopf point to find an unfolding which gives qualitative information about the behaviour of the original model. The resulting single function is shown to have the form

$$G(x, \theta) = a (x^2, \theta) x = 0, \quad a (0, 0) = 0$$

where $x = 0$ for the steady state, and solutions of $a = 0$ for $x \neq 0$ representing periodic solutions for the original system. We simplify our analysis by defining $z = x^2$. The Hopf hypotheses (**H2**) and (**H3**) are found equivalent to $a_\theta \neq 0$ and $a_z \neq 0$, respectively, with $v' = a_\theta$ and $a_l = -a_z/a_\theta$. Now singularity theory can be used for analyzing the non-trivial solutions of the single scalar function a $(z, \theta) = 0$. The singularity theory provides a complete classification for all problems with codimension less than three. It provides all the perturbed bifurcation diagrams for Hopf theorem degeneracies in terms of transition sets, which are the union of five varieties defined by:

Bifurcation variety, $\mathcal{B}_0 = \{\alpha \in IR^m,: \exists \theta \in a = a_\theta = 0 \text{ at } (0, \theta, \alpha)\}$

Bifurcation variety, $\mathcal{B}_1 = \{\alpha \in IR^m,: \exists (z, \theta), z > 0 \in a = a_z = a_\theta = 0 \text{ at } (zz, \theta, \alpha)\}$

Hysteresis variety, $H_0 = \{\alpha \in IR^m,: \exists \theta \in a = a_z = 0 \text{ at } (0, \theta, \alpha)\}$

Hysteresis variety, $H_1 = \{\alpha \in IR^m,: \exists (z, \theta), z > 0 \in a = a_z = a_{zz} = 0 \text{ at } (0, \theta, \alpha)\}$

Double limit variety, $D = \{\alpha \in IR^m,: \exists \theta, z_1, z_2 (z_1 \neq z_2; z_1, z_2 \geq 0), \in a = za_z = 0$
$$\text{at } (z_1, \theta, \alpha) \text{ and } (z_2, \theta, \alpha)\}$$

The bifurcation associated with these transitions will be presented in the following sections. We use the notation for the derivatives of a

$$a_{ij} = \frac{\partial^{i+j} a}{\partial z^i \, \partial \theta^j}$$

in order to simplify the $\mathbf{H_{km}}$ degeneracies conditions. These conditions are expressed in term of the original function F instead of the function $a (x^2, \theta) x$ in Farr [1986] and Golubitsky and Langford [1981].

6.3.1 $\mathbf{H_{0m}}$

These degeneracies occur when the transversality hypothesis is violated. The defining conditions for this family in term of the single scalar equation, $a (z, \theta) = 0$, are

$$a_{01} = a_{02} = \ldots = a_{0m} = 0, \quad a_{10} \neq 0, \quad a_{0m+1} \neq 0 \qquad (6.12)$$

The normal forms and the universal unfoldings for the first two members of this family are given in Table 6.3. The transition varieties and local diagrams are given in Figure 6.3. The H_{01}, which is equivalent to the $\mathcal{B}_0$ set, is associated with degeneracies at the Hopf point $(z = 0)$. The H_{01} bifurcation describes the coalescence (or the appearance) of two Hopf points while the H_{02} involves the appearance of three non-degenerate Hopf points. So this family can be described qualitatively by saying that it is associated with the appearance of $m + 1$ non-degenerate Hopf points.

Bifurcation Type	Normal Form	Universal Unfolding	Nondegeneracy Conditions
H_{01}^{+}	$x(x^2 + \theta^2) = 0$	$x(x^2 + \theta^2 + \alpha_1) = 0$	$a_{10} > 0, \; a_{02} > 0$
H_{01}^{-}	$x(-x^2 + \theta^2) = 0$	$x(-x^2 + \theta^2 + \alpha_1) = 0$	$a_{10} < 0, \; a_{02} > 0$
H_{02}^{+}	$x(x^2 + \theta^3) = 0$	$x(x^2 + \theta^3 + \alpha_1 + \alpha_2\theta) = 0$	$a_{10} > 0, \; a_{03} > 0$
H_{02}^{-}	$x(x^2 - \theta^3) = 0$	$x(x^2 - \theta^3 + \alpha_1 + \alpha_2\theta) = 0$	$a_{10} < 0, \; a_{03} > 0$

Table 6.3. Normal forms and unfoldings for H_{01} and H_{02}.

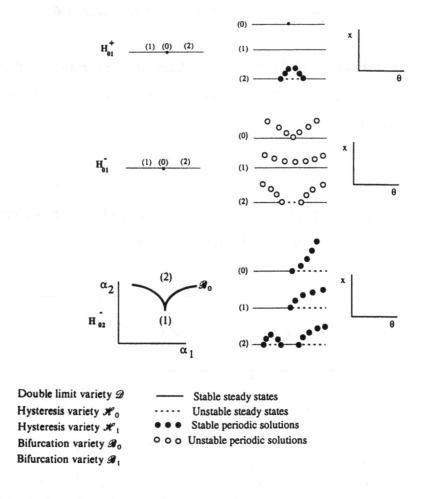

Double limit variety $\mathscr{D}$ —— Stable steady states
Hysteresis variety $\mathscr{H}_0$ ····· Unstable steady states
Hysteresis variety $\mathscr{H}_1$ ●●● Stable periodic solutions
Bifurcation variety $\mathscr{B}_0$ ○○○ Unstable periodic solutions
Bifurcation variety $\mathscr{B}_1$

Figure 6.3. Bifurcation diagrams for H_{01} and H_{02} (after Golubitsky and Schaeffer, [1985] with permission).

6.3.2 H_{k0}

H_{k0} degenerate Hopf points require the conditions

$$a_{10} = a_{20} = \ldots = a_{k0} = 0, \quad a_{01} \neq 0, \, a_{k+1,0} \neq 0$$

The normal forms and universal unfoldings of the first two members of this family are given in Table 6.4. The corresponding local bifurcation diagrams are shown in Figure 6.4. The $\mathbf{H}_{10}$ bifurcation is associated with the transition set H_0 defined at the Hopf point. It describes transition from supercritical to subcritical Hopf bifurcation. This transition leads to the formation of a limit or turning point in the periodic branch. The second kind $\mathbf{H}_{20}$ allows individual Hopf point to produce multiple periodic orbits. The parameter space of this kind includes two more transition boundaries. The transition across the set H_1 leads to the formation of hysteresis loops in the periodic branch, while the transition across the double limit variety D changes the relative positions of the Hopf point and the periodic branch limit point.

Bifurcation Type	Normal Form	Universal Unfolding	Nondegeneracy Conditions
$\mathbf{H}_{10}^{+}$	$x(x^4 + \theta) = 0$	$x(x^4 + \theta + \alpha_1 x^2) = 0$	$a_{10} > 0,\ a_{20} > 0$
$\mathbf{H}_{10}^{-}$	$x(x^4 - \theta) = 0$	$x(x^4 - \theta + \alpha_1 x^2) = 0$	$a_{01} < 0,\ a_{20} > 0$
$\mathbf{H}_{20}^{+}$	$x(x^6 + \theta) = 0$	$x(x^6 + \theta + \alpha_1 x^2 + \alpha_2 x^4) = 0$	$a_{01} > 0,\ a_{30} > 0$
$\mathbf{H}_{20}^{-}$	$x(x^6 - \theta) = 0$	$x(x^2 - \theta^3 + \alpha_1 + \alpha_2 \theta) = 0$	$a_{01} < 0,\ a_{30} > 0$

Table 6.4. Normal forms and unfoldings for $\mathbf{H}_{10}$ and $\mathbf{H}_{20}$.

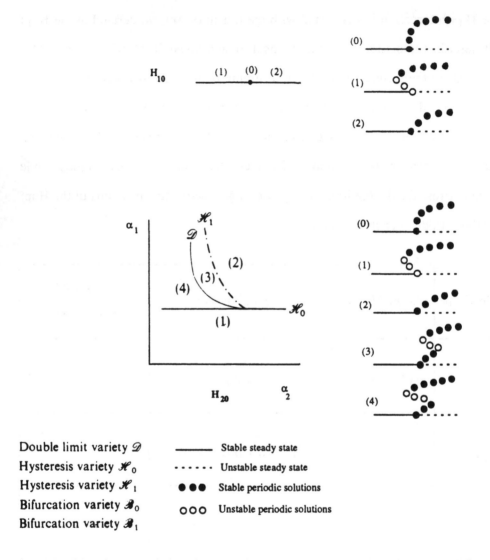

Double limit variety $\mathscr{D}$ ——— Stable steady state

Hysteresis variety $\mathscr{H}_0$ ······ Unstable steady state

Hysteresis variety $\mathscr{H}_1$ ●●● Stable periodic solutions

Bifurcation variety $\mathscr{B}_0$ ○○○ Unstable periodic solutions

Bifurcation variety $\mathscr{B}_1$

Figure 6.4. Bifurcation diagrams for H_{10} and H_{20} (after Golubitsky and Schaeffer, [1985] with permission).

6.3.3 H_{11}

Now we consider the simplest type of the third family, H_{11}, where both the transversality and the stability hypotheses fail. This is not a unique designation because there are three distinct non-degenerate members and only one name. The defining conditions are

$$a_{10} = a_{01} = 0, \quad a_{20} \neq 0, \quad a_{02} \neq 0 \tag{6.14}$$

The normal form has the form

$$x\,(\,\varepsilon x^4 + 2m\,\theta x^2 + \gamma\theta\,) = 0 \tag{6.15}$$

where $\varepsilon = sign\,(a_{20})$, $\gamma = sign\,(_{02})$ and $m = a_{11}/2\sqrt{|a_{20}a_{02}|}$, as long as $m^2 \neq \varepsilon\gamma$. The parameter m is called the modal parameter, and the fact that it appears in the normal form is the first clue to its difference from the parameters that appear in the unfolding. Usually the coefficients in the normal form are scaled to $\pm\,1$, but they cannot be in this case. The mathematical ideas behind moduli are too involved to go into here, but we will try to describe the structure of bifurcations involved with this family .

The universal unfolding of equation 6.15 is

$$x\,(\,\varepsilon x^4 + 2m\,\theta x^2 + \gamma\theta + \alpha_1 + 2\alpha_2 x^2\,) = 0$$

There are three distinct bifurcation associated with equaiton 6.16 depending on ε, γ, and m. The first occurs if $\varepsilon\gamma = -\,1$. The transition sets and diagrams are shown in Figure 6.5. This figure includes the set β_1 which is associated with transcritical bifurcation of

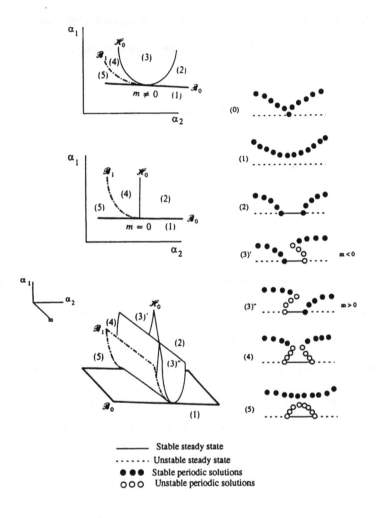

Figure 6.5. Bifurcation diagrams for H_{11}, $\varepsilon = 1$, $\gamma = -1$ (after Golubitsky and Schaeffer, [1985] with permission).

periodic orbits. The second set of bifurcation diagrams corresponds to $\varepsilon = \gamma = 1$ and $|m| > 1$. The transition set and the bifurcation diagrams are shown in Figure 6.6. The diagrams are exactly the same for $\varepsilon = \gamma = -1$ and $m < -1$, but the signs of α_1 and α_2 must be changed. The final bifurcation diagrams associated with equation 6.16 is for the case $\varepsilon\gamma = 1$ and $|m| < 1$. The transition set and the bifurcation diagrams with $\varepsilon = \gamma = 1$ are shown in Figure 6.7. It can be seen that isolated periodic solutions occur close to this kind of $\mathbf{H_{11}}$.

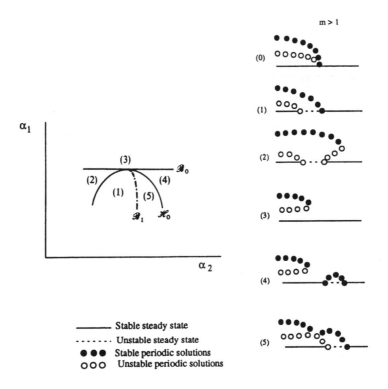

Figure 6.6. Bifurcation diagrams for $\mathbf{H_{11}}$, $\varepsilon = \gamma = 1, |m| > 1$ (after Golubitsky and Schaeffer, [1985] with permission).

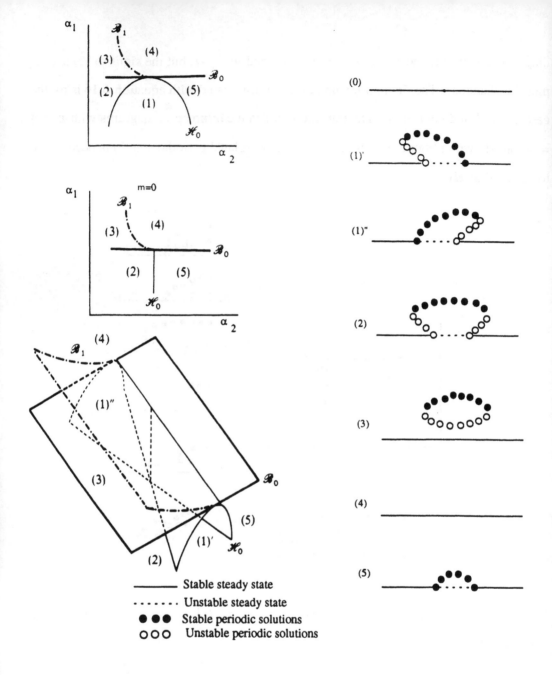

Figure 6.7. Bifurcation diagrams for $\mathbf{H}_{11}$, $\varepsilon = \gamma = 1$, $|m| < 1$ (after Golubitsky and Schaeffer, [1985] with permission).

6.4   Secondary Bifurcations

The concept of Poincaré maps will be used to define the mechanisms by which periodic solutions lose their stability and lead to secondary bifurcations [Guckenheimer and Holmes, 1983]. The periodic orbit γ with γ a period $T > 0$ is a set of points $x \in \gamma$ for which some mapping $g^{t+T}(x) = g^t(x)$ and $g^{t+s} \neq g^t(x)$ for all $0 < s < T$. Each point of the periodic orbit is a fixed point of the mapping g^T which can be represented by a discrete function. We consider the autonomous system

$$F(x, \theta) = 0 \tag{6.17}$$

To study the stability of the periodic orbit, it is advantageous to linearize the continuous dynamical system along the orbit γ which leads to non-autonomous system of linear equations with T-periodic coefficients

$$\dot{\varsigma} = A(\tau)\,\varsigma \tag{6.18}$$

where the matrix

$$A(t) = D_x F\left(x(t), \theta\right)\big|_{x(t) \in \gamma}$$

is periodic in t, $A(t+T) = A(t)$. It is proved in the theory of differential equations with periodic coefficients, Floquet theory [Guckenheimer, and Holmes, 1983], that any fundamental solution matrix for such periodic problems can be written in the form

$$U(t) = z(t)e^{tR},$$

where R is a constant matrix and $Z(t)$ is a periodic matrix of period T. If we choose

$$U(0) \ = \ Z(0) \ = \ I$$

then we obtain the monodromy matrix

$$U(T) \ = \ Z(T)e^{TR} \ = \ Z(0)e^{TR} \ = \ e^{TR}$$

The behaviour of solutions in the neighborhood of γ is determined by the eigenvalues of the constant matrix e^{TR}. Those eigenvalues are called the characteristic (Floquet) multipliers or roots. The multiplier associated with perturbations along γ is always unity. If all of the remaining multipliers of the monodromy matrix have moduli less than one, then γ is stable, otherwise the periodic orbit is unstable. A periodic orbit which does not have any eigenvalue of modulus equal to one is called hyperbolic. Poincaré mapping reduce the dimension of the dynamic system by defining the equivalent discrete system as following. First, we choose in the neighborhood of the point $x_o \in \gamma$ a surface Σ transverse to γ at the point x_0. Orbits in the vicinity of γ will intersect Σ close to the point x_0 (Figure 6.8). By following a sequence of intersections $\{x_k\}$ of an orbit close to γ with Σ we define the mapping $x \rightarrow P(x)$ from Σ into itself.

The stability of the orbit is determined by studying the linearized mapping $D_x P$. If we choose the basis appropriately so the last column of e^{TR} is $(0, ..., 0, 1)^T$, the matrix $DP(x)$ is simply the $(n - 1) \times (n - 1)$ matrix obtained by deleting the nth row and column of e^{TR}. Therefore the eigenvalues of the linearized mapping are equal to the eigenvalues of the matrix $U(T) \ = \ e^{TR}$ except for the eigenvalue corresponding to the direction tangenial to the orbit, which is equal to one. It should be clear that the stability of

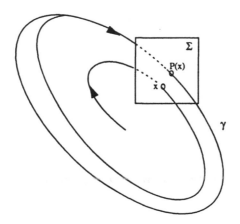

Figure 6.8. The Poincaré map.

periodic solutions is more difficult to determine, even when we reduce the system dimension by Poincaré mapping. This because the mapping function cannot be found explicitly unless the periodic solution is known analytically. This is seldom the case and numerical techniques need to be used [Doedel and Kerevez, 1981].

Local bifurcation in the neighborhood of the periodic orbit γ occurs when some eigenvalues of the linearized mapping lie on the unit circle in the complex plane. A saddle-node bifurcation, simple or transcritical bifurcation and pitchfork bifurcation are associated with one nontrivial eigenvalue equals one. They are completely analogous to the corresponding bifurcations of steady state, the unit circle taking the place of the imaginary axis. The conditions for these bifurcations are given in Table 6.5 [Wiggins, 1990]. We have seen that that saddle-node bifurcations of periodic solutions exist close to the degenerate Hopf points H_{10} (Figure 6.4). The simple or transcritical bifurcation are shown in the bifurcation diagrams obtained by perturbing the system around the H_{11} (Figure 6.5).

Bifurcation Type	Normal Form	Conditions
Saddle-node	$x + \theta \pm x^2$	$P = 0, P_x = 1, P_\theta \neq 0, P_{xx} \neq 0$
Simple	$x + \theta x \pm x^2$	$P = 0, P_x = 1, P_\theta = 0, P_{x\theta} \neq 0, P_{xx} \neq 0$
Pitchfork	$x + \theta x \pm x^3$	$P = 0, P_x = 1, P_\theta = 0, P_{xx} = 0, P_{x\theta} \neq 0, P_{xxx} \neq 0$

Table 6.5. Bifurcation of periodic solutions.

Another type of bifurcation, which has no analogy in static bifurcations, occurs when a single multiplier or eigenvalue passes through -1. In this case, a new system of periodic orbits with double period emanates from this bifurcation point (Figure 6.9). The map P requires the following conditions to have period-doubling bifurcation:

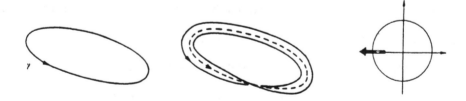

Figure 6.9. Period-doubling bifurcation.

120

$$P = 0, \ P_x = -1, \ P_\theta = 0, \ P_{xx} \neq 0, \ P_{x\theta} \neq 0$$

The normal form for period doubling bifurcation is

$$x \rightarrow P(x) = -x - \theta x + x^3 = 0$$

It has been shown that the generic bifurcations of fixed points of one-parameter dimensional maps are saddle-node and period-doubling.

The case of two complex conjugate eigenvalues passing through the unit circle is analogous to the Hopf bifurcation of the equilibrium points, however this case is more complicated. If two complex conjugate eigenvalues pass through the unit circle with a nonzero speed at the bifurcation point, then in the neighborhood of the bifurcation point, a system of two-dimensional tori arises. This bifurcation is known as "Naimark-Sacker bifurcation". As in the case of the Hopf bifurcation, the dimension of the formed object is higher by one. The bifurcated tori can contain either quasiperiodic orbits which cover the torus densely or periodic orbits based on the following. At the torus bifurcation, the system develops a second characteristic frequency, if the original system frequency and this frequency are rational, the resulting motion leads to a trajectory that closes up and does not cover the the complete surface of the torus (Figure 6.10(a)). If the frequencies are incommensurate (irrational), the combined effect will be to cause the system to fill the complete two-dimensional surface of the torus (Figure 6.10(b)).

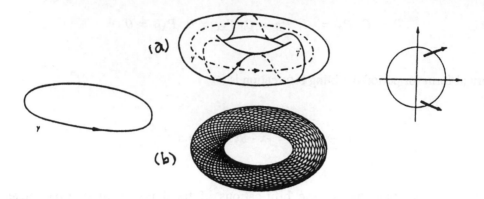

Figure 6.10. Torus bifurcation.

Chapter 7

DYNAMICS - MODEL I

The dynamic behaviour of the first model is considered in this chapter. First we present the conditions for Hopf points in terms of the characteristic equation coefficients. In the second section we examine the degeneracies obtained when the first hypotheses of Hopf theorem is violated. In Section 7.3 we present all the perturbed bifurcation diagrams for type-two degeneracies of codimension $\leq$ 2. In the last section we show the bifurcation diagrams which are associated with breaking the stability condition of Hopf Theorem.

7.1 Hopf Bifurcation

The mathematical model for the first case, coupling through the reactant, is:

$$\frac{dx}{d\tau} = \frac{z - x}{\lambda} - \frac{x}{\theta} - xy^2$$

$$\frac{dy}{d\tau} = \frac{\beta - y}{\theta} + xy^2 - \kappa y$$

$$\rho \frac{dz}{d\tau} = -\frac{z - x}{\lambda} + \frac{\sigma}{\theta}(1 - z) \qquad (7.1)$$

This three-dimensional system has a Hopf bifurcation when the Jacobian matrix has pure imaginary eigenvalues and the third eigenvalue is negative. The Jacobian matrix for this model is:

$$J = \begin{pmatrix} -\dfrac{1}{\lambda} - \dfrac{1}{\theta} - y_{ss}^2 & -2x_{ss}y_{ss} & \dfrac{1}{\lambda} \\[2ex] y_{ss}^2 & -\dfrac{1}{\theta} + 2x_{ss}y_{ss} - \kappa & 0 \\[2ex] \dfrac{1}{\rho\lambda} & 0 & -\dfrac{1}{\rho\lambda} - \dfrac{\sigma}{\rho\theta} \end{pmatrix} \qquad (7.2)$$

The eigenvalues μ of the Jacobian matrix are the solutions of the characteristic matrix equation:

$$-\mu^3 + S_1\mu^2 - S_2\mu + S_3 = 0 \qquad (7.3)$$

where S_1, S_2, and S_3 are the three invariants of J

$$S_1 = j_{11} + j_{12} + j_{33} \qquad (7.4)$$

$$S_2 = \det\begin{pmatrix} j_{11} & j_{12} \\ j_{21} & j_{22} \end{pmatrix} + \det\begin{pmatrix} j_{22} & j_{23} \\ j_{32} & j_{33} \end{pmatrix} + \det\begin{pmatrix} j_{11} & j_{13} \\ j_{31} & j_{33} \end{pmatrix} \qquad (7.5)$$

$$S_3 = \det(J) \qquad (7.6)$$

where the terms $j_{11}, j_{12}, \ldots$, are the elements of J. The condition for Hopf bifurcation in term of the coefficients S_1, S_2, and S_3 can be simply derived by setting $\mu = v + i\omega$ into equation 7.3. We obtain two equations from the real and imaginary parts:

$$\omega^2(3v - S_1) = v^3 - S_1v^2 + S_2v - S_3 \qquad (7.7)$$

$$\pm\omega(\omega^2 - 3v^2 + 2S_1v - S_2) = 0 \qquad (7.8)$$

124

he last equation gives ω^2:

$$\omega^2 = 3v^2 - 2S_1 v + S_2 \qquad (7.9)$$

Substitution of this result into equation 7.7 gives

$$8v^3 - 8S_1 v^2 + 2(S_1^2 + S_2)v = S_1 S_2 - S_3 \qquad (7.10)$$

We obtain general conditions for Hopf bifurcation for three dimensional system by setting $v = 0$ into the relations equation 7.10 and equation 7.9:

$$F = S_1 S_2 - S_3 = 0 \qquad (7.11)$$

$$\omega^2 = S_2 > 0 \qquad (7.12)$$

For the first model the condition F can be easily shown to be quartic in x_{ss} which means that the maximum number of simple Hopf points in any bifurcation diagram is four. We use the steady state equations and equation 7.11 together to solve for the x_{ss}, y_{ss}, and θ (or any other parameter). For example, for the parameters $\beta = .15$, $\sigma = 1$, $\lambda = 0.04$, $\rho = 5$, Figure 7.1 shows curve of Hopf points in the θ, κ-plane. With these parameter values, a unique steady state solution can be found (Figure 7.13). In Figure 7.1, four regions can be distinguished. Region 1 has no Hopf points which means that the model exhibits stable unique solutions for each θ and κ belong to this region. In Regions 2 and 4, two Hopf points appear in the bifurcation diagram Therefore, the steady state curve is divided into three regions; two of them are stable. In Region 3, four Hopf points appear in the bifurcation diagram which divide the equilibrium solutions into five parts; three

125

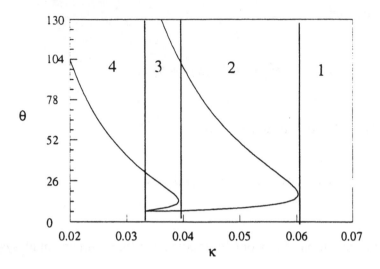

Figure 7.1. Hopf bifurcation points curve ($\beta = 0.15$, $\sigma = 1$, $\lambda = 0.04$, $\rho = 5$).

stable regions and two unstable regions. The bifurcation diagrams that correspond to these regions are shown in Figure 7.2. The appearance/disappearance of Hopf points is related to violating the transversality condition of Hopf theorem.

Next we illustrate the importance of the dynamic bifurcations by looking at the behaviour of the system when $\kappa = 0.0375$ in Figure 7.1. For this case the bifurcation diagram shows four Hopf points (Figure 7.3). The simulations of this model around the first three Hopf points ($\theta = 6.97, 9.50, 20.11$) are displayed in Figure 7.4. The model response changes dramatically when we run the model around these Hopf points. In Figure 7.5 we present the second example by using the same residence time ($\theta = 118$) but with different initial conditions. Here, the model has three solutions; one stable equilibrium solution, stable periodic solution and unstable periodic solutions (Figure 7.5a). It can be seen from this example that the number of the solutions of this model hanges when the periodic solution changes its stability or in other words when the third

ypothesis of Hopf theorem is not satisfied.

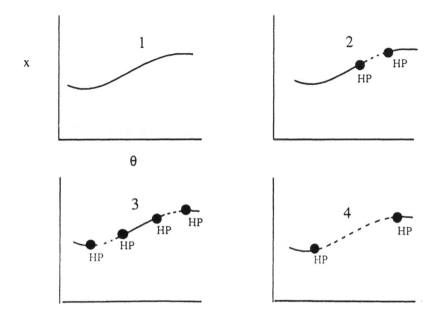

Figure 7.2. Bifurcation diagrams for the four regions identified in Figure 7.1.

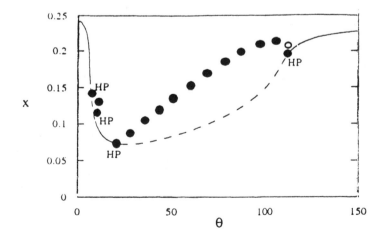

Figure 7.3. Bifurcation diagram ($\beta = 0.15$, $\kappa = 0.0375$, $\sigma = 1$, $\lambda = 0.04$, $\rho = 5$).

127

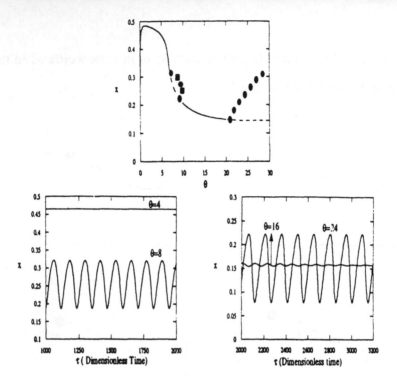

Figure 7.4. (a) Part of the bifurcation diagram shown in Figure 7.3. (b) and (c) Model simulation.

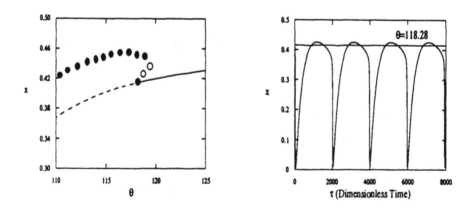

Figure 7.5. (a) Part of the bifurcation diagram shown in Figure 7.3. (b) Model simulation.

7.2   Degeneracies - Type I

7.2.1 F_1: Double-zero Eigenvalues

This degeneracy corresponds to the simplest interactions between Hopf point and a limit or turning point (static bifurcation). Here the imaginary part of the complex-conjugate eigenvalue pair associated with the Hopf point goes to zero as the Hopf and turning point come together, resulting in a double zero eigenvalue with Jordan block of size two:

$$\begin{pmatrix} 0 & 1 \\ 0 & 0 \end{pmatrix}$$

This Jordan block form means that there is only one eigenvector corresponding to the two zero eigenvalues. We will show that this is the case for our model. If there were two independent eigenvectors with eigenvalue zero, then the rank of the Jacobian matrix would be one. If we can construct a 2 x 2 matrix from the Jacobian (equation 7.2) such that it has nonzero determinant then we guarantee that our model does not have Jordan block of size one. Consider the following 2 x 2 matrix

$$\begin{pmatrix} j_{11} & j_{13} \\ j_{31} & j_{33} \end{pmatrix} = \begin{pmatrix} -\dfrac{1}{\lambda} - \dfrac{1}{\theta} - y_{ss}^2 & \dfrac{1}{\lambda} \\ \dfrac{1}{\rho\lambda} & -\dfrac{1}{\rho\lambda} - \dfrac{1}{\rho\lambda} \end{pmatrix}$$

The determinant of this matrix is:

$$\left(-\frac{1}{\lambda} - \frac{1}{\theta} - y_{ss}^2 \right)\left(-\frac{\sigma}{\rho\theta} \right) + \left(-\frac{1}{\theta} - y_{ss}^2 \right)\left(-\frac{1}{\rho\lambda} \right)$$

which is never equal to zero for positive parameters.

The F_1 configuration can be found by substituting $\mu_1 = 0$ and $\mu_2 = 0$ into the characteristic equation (equation 7.3) to get the following general conditions for this degenerate bifurcation:

$$S_2 = S_3 = 0 \qquad\qquad (7.13)$$

The conditions (equation 7.13) and the steady state equations can be solved simultaneously for the steady states and two parameters (say θ and β) in terms of the remaining parameters. We will fix the parameters σ, λ, and ρ and vary the decay parameter κ to define the boundaries in the κ, $\beta-$ plane at which these degeneracies occur. When crossing these boundaries, the number of Hopf points in the bifurcation diagrams increases/decreases by one. For example, consider the branch set shown in Figure 7.6, we can see from the bifurcation diagrams that when we cross the F_1 curve, the bifurcation diagram loses/gains one Hopf point. It is clear that F_1 degeneracies occur only in the regions where the bifurcation diagrams have turning points, i.e., the multiplicity region. Another comment to maintain is that F_1 degeneracy represents only one route for which a change in the number of the Hopf points can occur. Therefore studying these degeneracies alone, without including the other routes, is not enough to divide the branch set into regions with different number of Hopf points.

To examine the dynamic behaviour when perturbing the model around the F_1 degeneracies we consider the case shown in Figure 7.6, $\beta = .1125$, two F_1 singularities are found in the θ, $\kappa-$ plane (Figure 7.7). Homoclinc orbits can be found approximately by computing two-parameter continuation of limit cycles of large period [Doedel and Kerevez, 1983]. Figure 7.7 shows the loci of the turning points, Hopf points and homoclinc orbits around these two F_1 singularities. It shows also the bifurcation diagrams which are resulted from perturbing the model around these points.

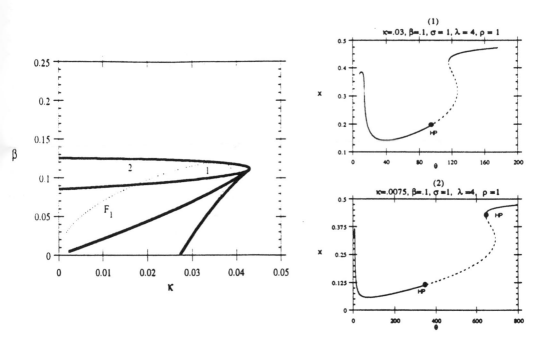

Figure 7.6. Branch set ($\sigma = 1, \lambda = 4, \rho = 1$).

7.2.2 F_2: Pure Imaginary Pair and a Simple Zero Eigenvalue

In this case, the Hopf point remains as a Hopf point as it passes through the limit point so that three eigenvalues are at criticality:

$$\begin{pmatrix} 0 & 0 & 0 \\ 0 & 0 & \omega \\ 0 & -\omega & 0 \end{pmatrix}$$

131

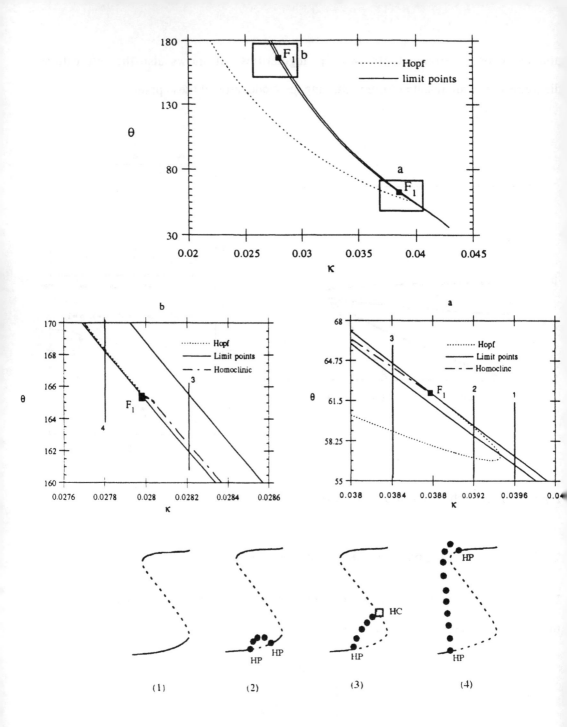

Figure 7.7. F_1 degeneracies in the θ, κ-plane.

The conditions for this degeneracy in terms of the characteristic coefficients are:

$$S_1 = 0 \qquad (7.14)$$

$$S_3 = 0 \qquad (7.15)$$

$$S_2 > 0 \qquad (7.16)$$

Suppose that the conditions (equation 7.14, equation 7.15) are fulfilled for some parameters. Then we need to check the sign of S_2 which takes the form:

$$\det \begin{pmatrix} j_{11} & j_{12} \\ j_{21} & j_{22} \end{pmatrix} + \det \begin{pmatrix} j_{22} & j_{23} \\ j_{32} & j_{33} \end{pmatrix} + \det \begin{pmatrix} j_{11} & j_{13} \\ j_{31} & j_{33} \end{pmatrix}$$

$$= j_{11}j_{22} - j_{12}j_{21} + j_{22}j_{33} + j_{11}j_{33} - j_{13}j_{31}$$

Note that the quantities j_{11}, j_{12} and j_{33} are always negative while j_{13}, j_{21}, and j_{31} are always positive for positive parameter values. j_{23} and j_{32} are zero.

The first condition $S_1 = 0$ can be written as

$$j_{11} + j_{22} = -j_{33}$$

which implies that $j_{22} > 0$. We substitute this expression into S_2 to get

$$S_2 = j_{11}j_{22} - j_{12}j_{21} - j_{33}^2 - j_{13}j_{31}$$

Multiplying this condition by the quantity j_{33} results in

133

$$j_{33}S_2 = j_{11}j_{22}j_{33} - j_{12}j_{21}j_{33} - j^3_{33} - j_{13}j_{31}j_{33}$$

Since $j_{33} < 0$, the condition $S_2 > 0$ is equivalent to $j_{33}S_2 < 0$. Now we use the second condition $S_3 = 0$ which takes the form:

$$S_3 = j_{11}j_{22}j_{33} - j_{12}j_{21}j_{33} - j_{13}j_{22}j_{31} = 0$$

to simplify the condition $j_{33}S_2$

$$j_{33}S_2 = j_{13}j_{22}j_{31} - j^3_{33} - j_{13}j_{31}j_{33}$$

Clearly $j_{33}S_2$ is always positive, (i.e., $S_2 < 0$) which means that $\mathbf{F_2}$ bifurcation does not occur for this model.

In a similar way, one can start by assuming that two of the $\mathbf{F_2}$ conditions, equations 7.14 - 7.16 hold, then he can show that the third condition is not satisfied. If we assume that $S_3 = 0$ and $S_2 \leq 0$, then it can be proven that $S_1 < 0$ for all positive parameter values. The $S_1 < 0$ means that the trace of the Jacobian matrix is always negative which means that the first model is dissipative at the singularities which satisfy $S_3 = 0$ and $S_2 \geq 0$. For a dissipative system, the dimensionality of any attractor must be less than the dimension of the full state of model. For such system, the long term behaviour cannot be quasi-periodic because quasi-periodic solutions live on the surface of a torus of dimension three. Thus, the condition $S_1 = 0$ is required for any model to exhibit stable torus bifurcation. This argument is based on the divergence theorem of dynamic systems which says roughly [Hilborn, 1994].

If div(f) $\equiv \sum_{i=1}^{N} \dfrac{\partial f_i}{\partial x_i} < 0$ on the average over state space, it is known that the initial volume of initial conditions will collapse onto geometric region whose dimensionality is less than that of original state space and the system is said to be dissipative.

According to this theorem, any model cannot exhibit quasi-periodic solutions if the model has $S_3 \equiv \text{div}(f) < 0$ throughout its state space.

7.2.3 G_1: Triple-zero Eigenvalues

The conditions for this degeneracy are:

$$S_1 = S_2 = S_3 = 0$$

It can be shown easily that G_1 does not occur in this model. We have shown in the last section that when the coefficients S_1, S_3 vanish, the third coefficient S_2 is always negative. This means that neither F_2 or G_1 exist for this model.

7.3 Degeneracies - Type II

This kind of degeneracies occurs when violating the transversality condition of Hopf theorem. The conditions for these degenerate Hopf points are given in Section 6.3.1. For system of low dimension it is possible to use the characteristic equation to define these degeneracies in the parameter space. The conditions for the type-2 degeneracies (H_{0m}) are

$$v = v' = v'' = \ldots = 0, \quad v^{(m+1)} \neq 0 \qquad (7.17)$$

where ν is the real part of the complex conjugate pair, and the derivatives are with respect to the bifurcation parameter (θ for this case). These conditions can be written in terms of the characteristic coefficients [Farr, 1986]:

$$F = F_\theta = F_{\theta\theta} = \ldots = \frac{d^m F}{d\theta^m} = 0, \quad \frac{d^{m+1} F}{d\theta^{m+1}} \neq 0 \tag{7.18}$$

where

$$F = S_1 S_2 - S_3 = 0, \quad S_2 > 0$$

Away from turning points, the implicit functional theorem can be used to define $x_{ss}(\theta)$ as an invertible function and this is sufficient to write the conditions of equation 7.18 as

$$F = F_x = F_{xx} = \ldots = \frac{d^m F}{dx^m} = 0, \quad \frac{d^{m+1} F}{dx^{m+1}} \neq 0 \tag{7.19}$$

The reason behind this that it is easier to express the derivatives with respect to the steady state than with respect to the parameter θ. Since F is quartic in x_{ss} it is possible to locate singularities where F and up to its first three derivatives vanish simultaneously, i.e., **H$_{01}$**, **H$_{02}$**, **H$_{03}$**. Locating these singularities is analogous to the procedure used for the steady state analysis.

7.3.1 H$_{01}$

This singularity corresponds to the appearance or coalescence of two Hopf points in the bifurcation diagrams. The conditions for **H$_{01}$** are:

$$F = F_x = 0, \quad F_{xx} \neq 0 \qquad (7.20)$$

where

$$F = S_1 S_2 - S_3 = 0, \quad S_2 > 0$$

This condition is equivalent to the simplest type of the static bifurcation, i.e., the turning points of the steady state curve in the bifurcation diagrams. H_{01} represents the turning points of the Hopf points curve. Note that the turning point of the steady state curve is of dimension zero, while H_{01} is of dimension one. This is because the zero set of F defined above already consists of Hopf points (H_{00}) which have codimension zero. Following the same procedure used in studying the codimension-one singularities (hysteresis, mushroom, isola, F_1), H_{01} can be represented in the branch set. We examine these singularities for seven cases.

Case 1: $\sigma = 1, \lambda = 4, \rho = 1$

First, we consider the case $\sigma = 1, \lambda = 4, \rho = 1$, we solve the steady state relations and the conditions in equation 7.20 for the steady state variables x_{ss}, y_{ss}, and the parameters θ, k in term of the remaining parameter β. This results in a unique curve for the H_{01} singularities in the branch set shown in Figure 7.8. This curve passes through the region where the model shows unique solutions. It can be seen from Figure 7.8 that the H_{01} loci starts at the point of the tangency between the F_1 curve and the mushroom-hysteresis curve. The branch set for these parameter values is qualitatively equivalent to the case of the Gray and Scott CSTR. The steady state locus exhibits Hopf bifurcation (dynamic instability) at some residence times θ for parameters κ and β which lie to the

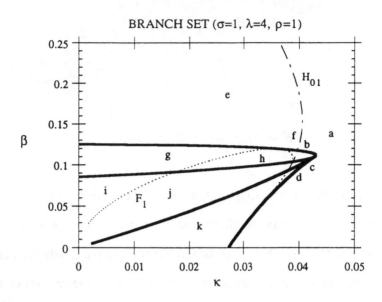

Figure 7.8. Branch set ($\sigma = 1, \lambda = 4, \rho = 1$).

left of or below the curve $\mathbf{H_{01}}$. Thus, Hopf bifurcation are favored by small values for the decay-rate constant, κ.. The parameter space can be divided into regions of different number of Hopf points. This allows us to give a complete classification with respect to the stability of the equilibrium points. Eleven regions are found for this case. No Hopf points appear in Regions a,b,c and d which are shown in Figure 7.9. In these regions the equilibrium points lose (or gain) their stability by static bifurcation (turning points). The bifurcation diagrams in Region e have unique solutions with two Hopf points. Regions f and g show hysteresis-shape equilibrium curve with two Hopf points. In Region f, the equilibrium points lose and gain their stability first by Hopf bifurcation, then they lose and gain their stability by static bifurcation. In Region g, the static bifurcations occurs between the two Hopf points. In Region i, we have mushroom-type equilibrium curve with two Hopf points. Because of crossing $\mathbf{F_1}$ degeneracies, the bifurcation diagrams in

138

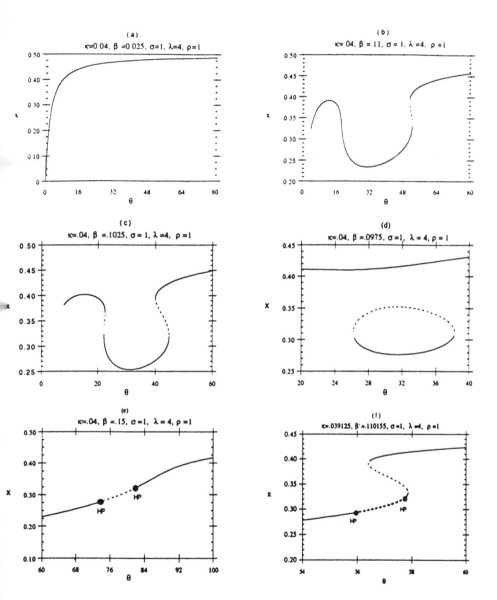

Figure 7.9. Bifurcation diagrams ($\sigma = 1$, $\lambda = 4$, $\rho = 1$).

Figure 7.9 (continued)

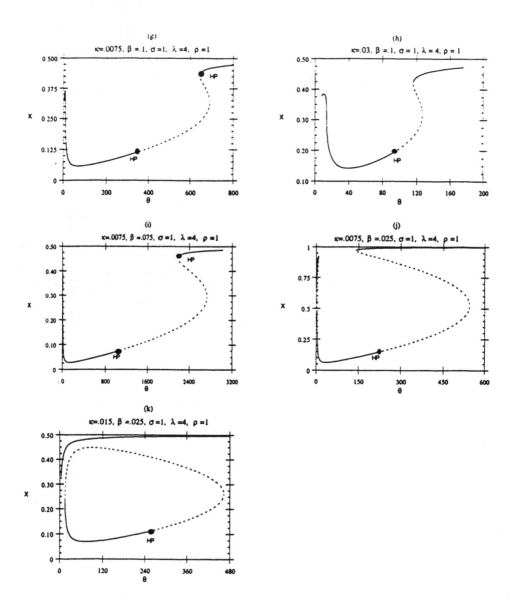

Regions h (hysteresis), j (mushroom) and k (isola) have only one Hopf point. Clearly, one can see that no regions with more than two Hopf points are found for this case. Since it is possible for the bifurcation diagrams in this model to have up to four Hopf points, we need to study these codimension-one singularities for different parameters sets.

Case 2: $\sigma = 1, \lambda = 4, \rho = 5$

Twelve distinct bifurcation diagrams are found for this case (Figure 7.10). The appearance of a second H_{01} curve inside the multiplicity region has resulted into eight

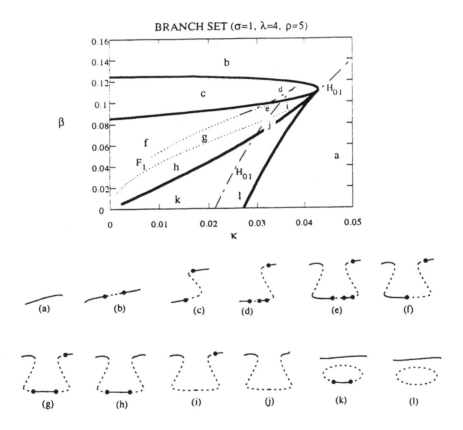

Figure 7.10. Branch set ($\sigma = 1, \lambda = 4, \rho = 5$).

new diagrams (d, e, g, h, i, j, k, l). This curve has a sharp turning point which is denoted by H_{02}. Inside the Regions d and e, we find hysteresis-shape and mushroom bifurcation diagrams having four Hopf points. Bifurcation diagrams of three Hopf points are also obtained inside the Region g. It can be seen that the mushroom-type equilibrium diagrams (h, i, and j) and the isola equilibrium diagrams (k and l) have less stability region than the corresponding regions for the case of a smaller tank ratio ($\rho = 1$), Figure 7.8.

Case 3: $\sigma = 1, \lambda = 4, \rho = 20$

Increasing the parameter ρ has resulted in enlarge the region of the κ, β–plane which is affected by the internal $\mathbf{H_{01}}$ curve (Figure 7.11). The external $\mathbf{H_{01}}$ curve coincides with the boundary at which the isola grows to a mushroom and starts at the point of tangency between the $\mathbf{F_1}$ curve and the isola-mushroom boundary. Three more new bifurcation diagrams are found. The first one is a unique solution with four Hopf points (c). The second one is given in f, a mushroom curve with four Hopf points. The third bifurcation diagram, j, shows a mushroom pattern with two Hopf points which lie close to the top turning points.

It can be seen from the previous cases that increasing the parameter ρ has resulted in creating more Hopf points in the bifurcation diagrams which decreases the stability regions of the equilibrium points of these diagrams. Figure 7.12 shows the effect of ρ on the first (external) $\mathbf{H_{01}}$ curve. It can be seen that increasing ρ enlarges the region of the dynamic bifurcation with respect to the parameters κ and β.

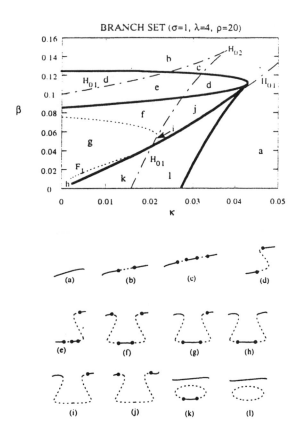

Figure 7.11. Branch set ($\sigma = 1, \lambda = 4, \rho = 20$).

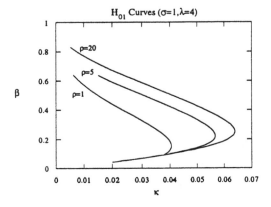

Figure 7.12. Effect of the parameter ρ on the external H_{01} curve ($\sigma = 1, \lambda = 4$).

143

Case 4: $\sigma = 1$, $\lambda = 0.04$, $\rho = 5$

The branch set for this case shows fourteen distinct regions (Figure 7.13). The bifurcation diagrams which correspond to these regions, are similar to those obtained for the previous cases.

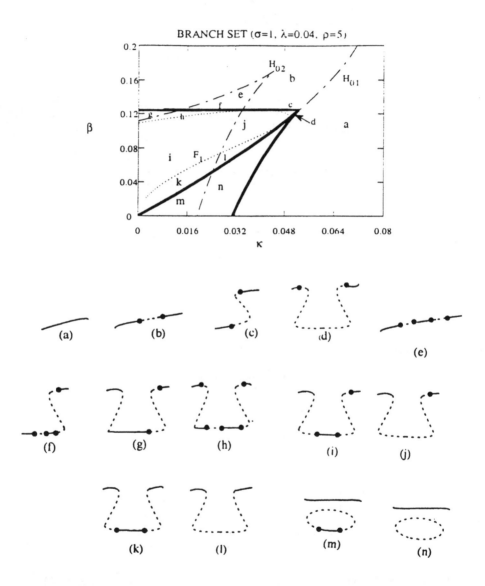

Figure 7.13. Branch set ($\sigma = 1$, $\lambda = 0.04$, $\rho = 5$).

Case 5: $\sigma = 1, \lambda = 400, \rho = 5$

Sixteen bifurcation diagrams have been identified for this case. Eleven of them can be seen clearly in the branch set given in Figure 7.14A. The small region around (k) is enlarged in Figure 7.14B where five small regions are defined. We see for the first time hysteresis-shape with three Hopf points in Region m.

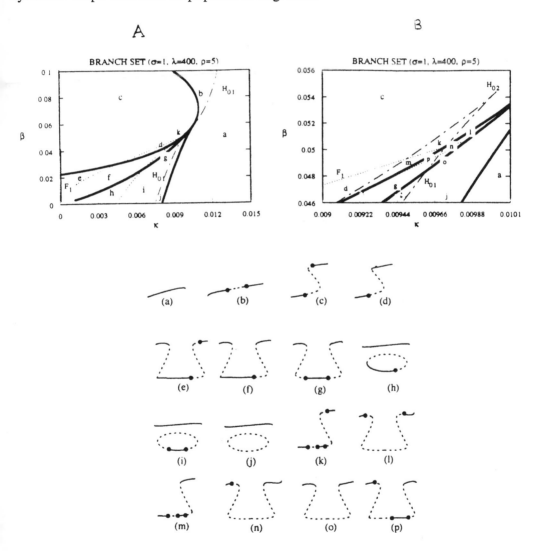

Figure 7.14. Branch set ($\sigma = 1, \lambda = 400, \rho = 5$).

145

By looking at the cases 2, 4, and 5, we find that decreasing the parameter λ increase the region affected by the internal H_{01} curve. The external H_{01} curves for these cases are plotted alone in Figure 7.15. It shows that increasing λ reduces the κ, β- Region where the dynamic or Hopf bifurcation may take place.

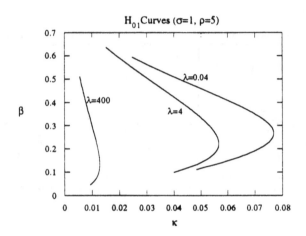

Figure 7.15. Effect of the parameter λ on the external H_{01} curve ($\sigma = 1$, $\rho = 5$)

Case 6: $\sigma = 0.5$, $\lambda = 9$, $\rho = 5$

For this case, the model shows fifteen bifurcation diagrams, (see Figure 7.16). We observe for the first time in Figure 7.16(e) an isola-equilibrium diagram with two Hopf points appearing not in the isolated solution but in the original equilibrium curve.

146

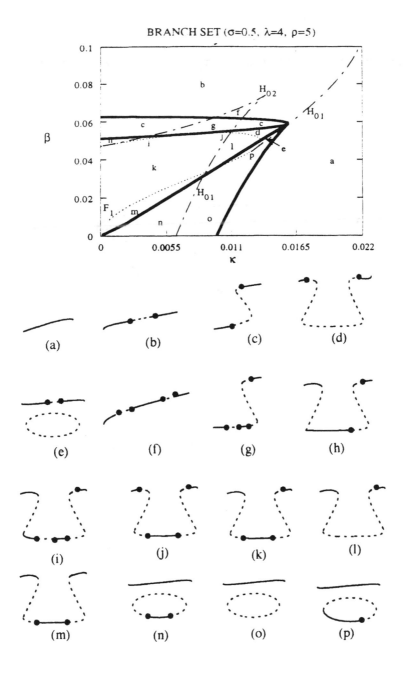

Figure 7.16. Branch set ($\sigma = 0.5$, $\lambda = 4$, $\rho = 5$).

147

Case 7: $\sigma = 2.0$, $\lambda = 2.25$, $\rho = 5$

Higher σ has resulted in fewer bifurcation diagrams; nine patterns have been found for this case, (see Figure 7.17).

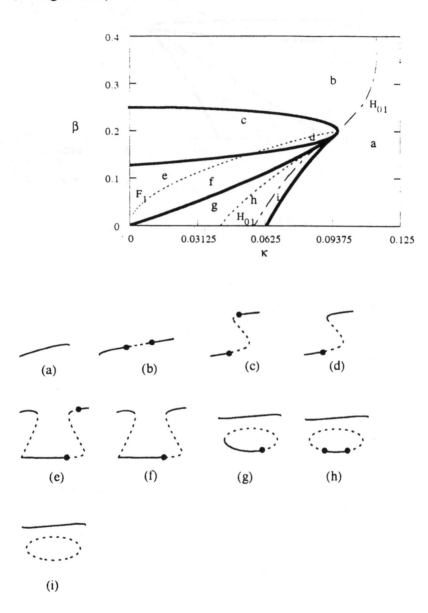

Figure 7.17. Branch set ($\sigma = 2.0$, $\lambda = 2.25$, $\rho = 5$)

From the cases 2, 6, and 7, we observe that the second or the internal H_{01} curve bounds more regions on the multiplicity region with lower σ. Figure 7.18 shows how the external H_{01} boundary looks for different values of σ. It can seen that the region where Hopf bifurcations occur has been enlarged by using larger flow-rate ratio parameter.

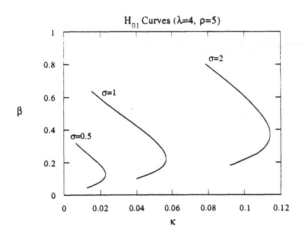

Figure 7.18. The effect of the paraemter σ on the external H_{01} curve ($\lambda = 4$, $\rho = 5$).

7.3.2 H_{02}

This singularity involves interactions between three Hopf bifurcation points. It is of codimension-two. The H_{02} degenerate Hopf points can be located simply from the following conditions:

$$F = F_x = F_{xx} = 0, \quad F_{xxx} \neq 0 \qquad (7.21)$$

149

where

$$FF = S_1 S_2 - S_3, \quad S_2 > 0$$

This singularity is analogous to the hysteresis singularity for the steady state multiplicity. We have found that the internal $\mathbf{H_{01}}$ curves which are obtained in some of the branch sets in Section 7.3.1, have turning points which satisfy the $\mathbf{H_{02}}$ conditions. In fact, these $\mathbf{H_{01}}$ curves have similar shape to the universal unfolding of the $\mathbf{H_{02}}$ singularity (see Section 6.3.1, Figure 6.3). To illustrate more, we examine the third case $\sigma = 1, \lambda = 4, \rho = 20$. Figure 7.19 shows part of the branch set where only a unique steady-state behaviour can be found. Three curves of Hopf points are plotted in Figure 7.20. It can be seen that the Hopf curves have different forms depending on whether the parameter β lies above or

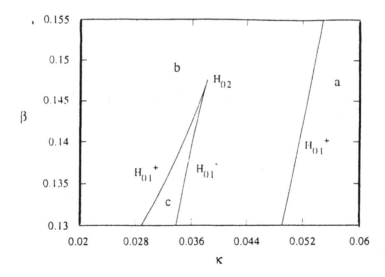

Figure 7.19. Branch set ($\sigma = 1, \lambda = 4, \rho = 20$).

150

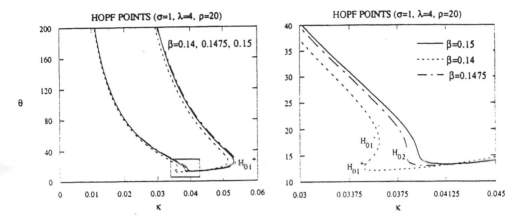

Figure 7.20. Hopf points curve.

below the H_{02} point. For β values above the H_{02} point, unique Hopf points curves exist in the θ, κ-plane, while for β values below the H_{02} point, multiple Hopf points curves are found.

We have seen in the last section that some of the branch sets do not have H_{02} singularities, i.e., the branch set has only one H_{01} curve (see, for example, the first case in the last section). Therefore, by studying H_{02} in the parameter space we can define the parameter values at which the number of H_{01} curves in the branch set changes. This tells us whether bifurcation diagrams of four Hopf points exist or not in the θ, κ–plane when fixing the remaining parameters. We use the conditions (Figure 7.21) together with the steady state relationships to solve for the variables x_{ss} and y_{ss} and the parameters θ, κ and β in terms of one of the remaining parameters. For fixed σ and ρ, a unique curve for the H_{02} points can be specified in the κ, λ and β, λ–parameters spaces. Figures 7.21, 7.22, and 7.23 show the loci of **H₀₂** for three values of σ (0.2, 1.0, 5.0) and three values of ρ (1.0, 5.0, 20.0). For low flow rate ratio (σ = 0.2), the model exhibits **H₀₂** degeneracies

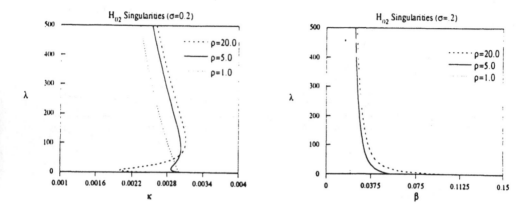

Figure 7.21. The **H₀₂** loci for σ = 0.2.

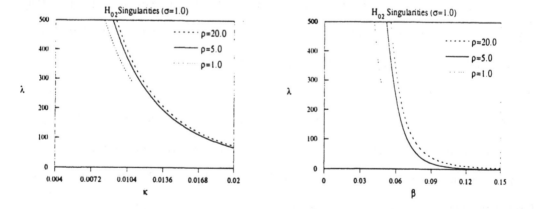

Figure 7.22. The **H₀₂** loci for σ = 1.0.

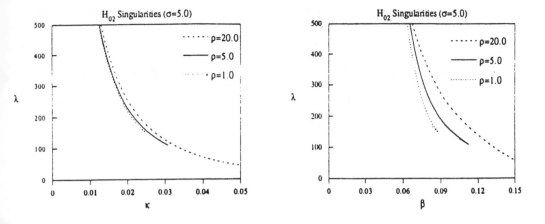

Figure 7.23. The $\mathbf{H_{02}}$ loci for $\sigma = 5.0$.

for all λ values, while for higher σ values the branch sets do not have $\mathbf{H_{02}}$ singularities for low λ. As an example, consider the $\mathbf{H_{02}}$ curve shown in Figure 7.22 which corresponds to the case $\sigma = 1, \rho = 1$. Clearly, no $\mathbf{H_{02}}$ points are defined for $\lambda < 280$. Similar observation can be found for the cases $\sigma = 5$ and $\rho = 1.0, 5.0$ which are plotted in Figure 7.23(a,b). These $\mathbf{H_{02}}$ curves can not extended because of violation of he Hopf inequality condition $S_2 > 0$. It can be also seen that when σ or ρ increases, the $\mathbf{H_{02}}$ singularity occurs at larger κ and β values, while increasing λ has the opposite effect.

7.4 Degeneracies - Type III

The third type of degeneracies allows individual Hopf points to produce multiple periodic orbits. It occurs when the third hypothesis of Hopf theorem is violated which means that one or more of the coefficients $(a_1(0), a_2(0),...)$ of the Hopf bifurcation normal form (equation 6.9) vanish simultaneously. The normal form of the Hopf point and the conditions and unfolding of the first two families H_{10}, H_{20} are given in Section 6.3.2. We will use the formulae derived by Golubitsky and Langford [1981] and modified by Farr [1986] to study the first two families of these degeneracies.

7.41 H_{10}

The first family of these degeneracies, referred as H_{01}, describes the condition for the change in stability of the emerging limit cycle. We follow this type of degeneracy as a curve across the parameter plane (κ, β) by using the formulas given by Golubitsky and Langford and modified by Farr [1986] (see also Golubitsky and Schaeffer [1985] together with the steady-state conditions and the Hopf bifurcation condition. Then we can combine these singularities with the other codimension-one degeneracies (steady-state singularities, F_1 and H_{01} together on the branch set. We examine the same cases studied in Section 7.3.1.

Case 1: $\sigma = 1, \lambda = 4, \rho = 5$

Two H_{01} loci are specified in the parameter plane (Figure 7.24). One of them emanates from the point of the tangency between the hysteresis and the F_1 curves. It passes through the region where the steady-state locus does not have multiplicity. At this boundaries, the Hopf point at the longer residence time changes its stability. It changes from a supercritical bifurcation (stable limit cycle emerging) for conditions to the right of the curve to subcritical (unstable limit cycle emerging) to the left. The other locus cuts through the parameter space for isola and mushroom patterns where the steady-state

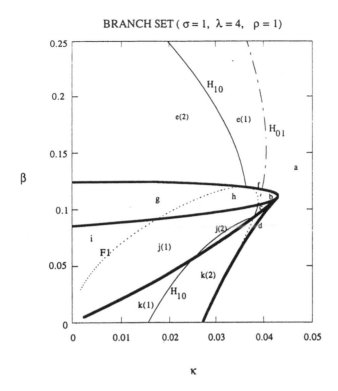

Figure 7.24. Branch set for $\sigma = 1, \lambda = 4, \rho = 1$

curve has only one Hopf point. It changes from a subcritical bifurcation for conditions to the right of the $\mathbf{H_{10}}$ to supercritical to the left. Fourteen regions are identified in Figure 7.24 by the various loci of steady-state and Hopf bifurcation degeneracies. Numerical values for the steady-state loci for these regions are given in Figure 7.25. In the bifurcation diagrams Figure 7.25 e(1), e(2), f, g, and i, the limit cycle emerging from one Hopf point grows until it collides with another limit cycles which has grown from another Hopf point. In cases h, j(1), j(2), k(1) and k(2), the limit cycles break down when they hit saddle-node points, where homoclinc orbits (HC) are obtained.

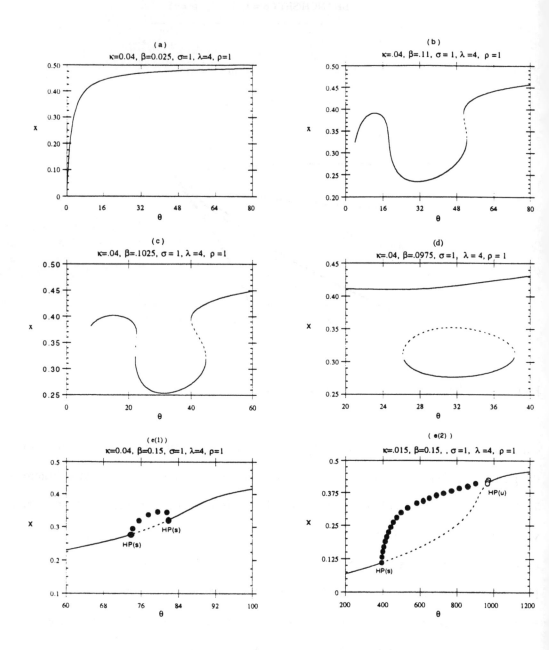

Figure 7.25. Bifurcation diagrams for case 1.

156

Figure 7.25 (continued)

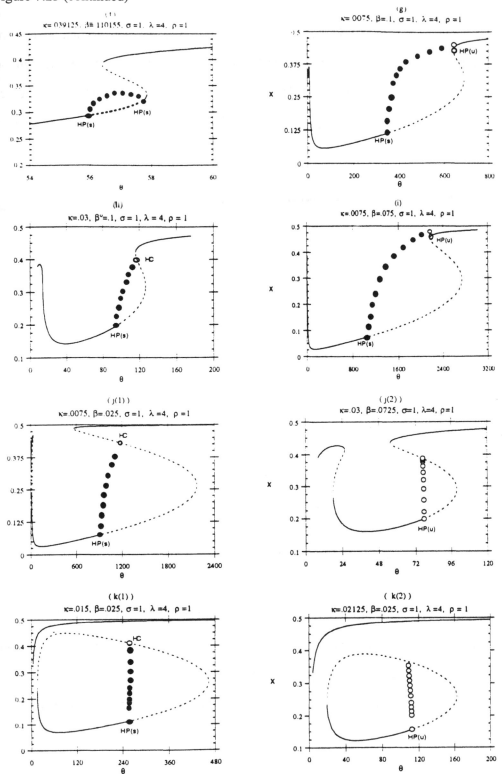

Case 2: $\sigma = 1, \lambda = 4, \rho = 5$

The branch set given in Figure 7.26 shows two H_{10} curves. The first starts from the point of tangency of the F_1 and the mushroom-hysteresis curve. It passes through the hysteresis and the unique regions. It also touches the H_{01} curve at a point where a higher order degeneracy is found, namely H_{11}. Around this point, three different unique steady-

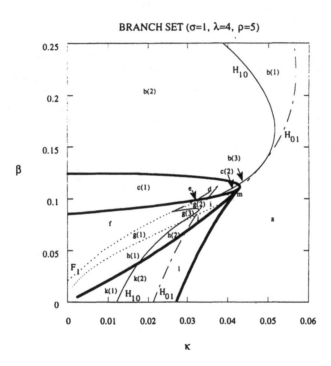

Figure 7.26. Branch set for $\sigma = 1, \lambda = 4, \rho = 5$.

state patterns can be identified in the Regions b(1), b(2) and b(3) shown in Figure 7.27. There are two Hopf points along the locus in these regions; in b(1), the bifurcation at both Hopf points are supercritical, in b(2), the bifurcation at longer residence time θ becomes subcritical, and in b(3), both Hopf bifurcation points are subcritical. This H_{10} curve

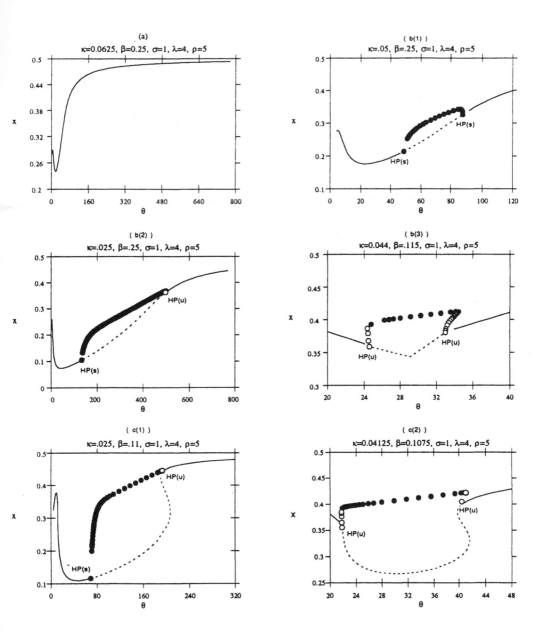

Figure 7.27. Bifurcation diagrams for case 2.

Figure 7.27 (continued)

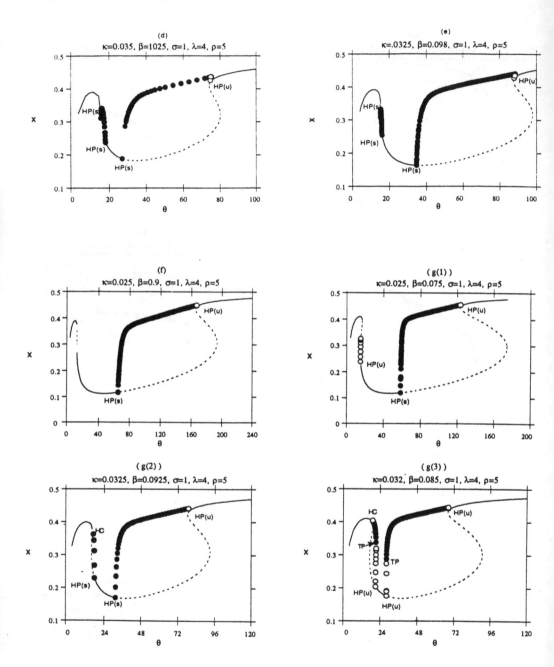

Figure 7.27 (continued)

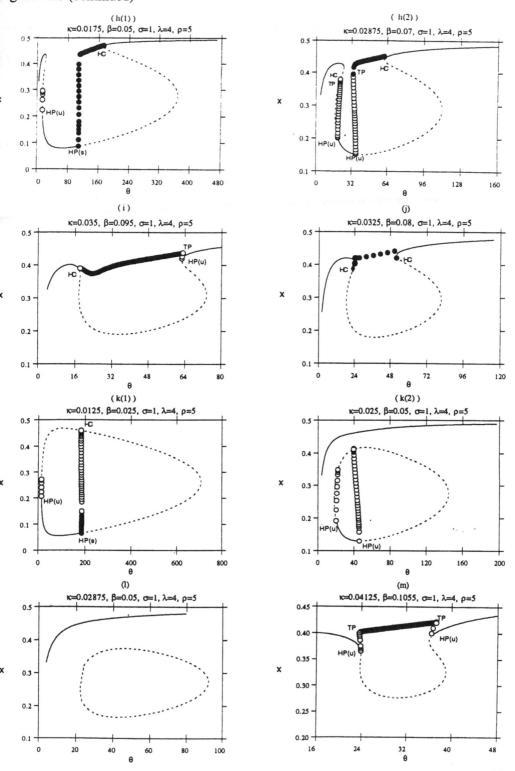

divides the hysteresis region into two parts. In c(1) the bifurcation diagram shows two Hopf points, a stable limit cycle emerge from the Hopf point at the smaller θ while an unstable limit cycle grows out of the second Hopf points. The steady state locus in c(2) has two subcritical Hopf points, i.e., unstable limit cycles emerge form both Hopf points. The other H_{10} curve cuts through the isola and mushroom regions. It has an influence on the stability of the two Hopf points appearing on the bottom curve of the mushroom and the isola patterns. The Regions g(1), g(2) and g(3) in Figure 7.26 show mushroom type steady-state loci which undergo three Hopf bifurcations. Two of them lie on the bottom part of the mushroom curve. In g(1), the first and the last Hopf point (i.e., at the smaller θ and the larger θ) are subcritical while the middle Hopf points is supercritical. If we cross the upper part of H_{10} curve going to g(2), the first Hopf point changes to supercritical while when we cross the H_{10} curve going to g(3), the middle Hopf point becomes subcritical. Note that these different bifurcation diagrams are found around the point of tangency of this H_{10} curve and the internal H_{01} curve. In the the Regions h(1) and h(2), as in g(1)and g(3), the stability of the limit cycles emerging from the second Hopf point changes when we cross the H_{10} curve. Similar changes are also observed with the Regions k(1) and k(2). Figure 7.27 gives numerical values for the bifurcation diagrams for the twenty regions obtained for this case. The periodic branches emanating from some of the Hopf points in Figure 7.27 exhibit the simplest kind of the secondary dynamic bifurcation; turning point (TP) or saddle-node bifurcation. The model around these Hopf points has three attractors: stable (unstable) equilibrium point, stable limit cycle, unstable limit cycle. One of the most interesting bifurcation diagrams is shown in Figure 7.27(j). There are no Hopf points in the equilibrium patterns of this region but there is periodic solution branch for some θ values at which all the equilibrium points are unstable. Here, the limit cycles in both sides of the branch die at saddle-node points

162

Homoclinic orbit). Another point to notice is that up to five attractors can be seen in some of the bifurcation diagrams of this case. For example, close to the residence time values which lie to the right of the second Hopf point of the S-shaped diagram shown in Figure 7.27c(2), there are three unstable equilibrium points, one stable limit cycle and unstable limit cycle. Similar configurations can be observed in the bifurcation diagrams of the Regions c(1), d, e, k(1) and m.

Case 3: $\sigma = 1, \lambda = 4, \rho = 20$

The branch set for this case is illustrated in Figure 7.28. Qualitative forms of the bifurcation diagrams for each region are given in Table 7.1. The H_{10} loci become more complicated. It can be seen that the more interesting regions are those lying around the points of tangency between the H_{10} and H_{01} curves or the points of the tangency or intersection of the H_{10} curves. There are two H_{10} loci for this case. One of them passes through the region of uniqueness and through the hysteresis region. It has similar effect to the effect of the first H_{10} locus which was examined in the last case. The other curve cuts through the whole multiplicity region. Its effect on the mushroom and the isola patterns are equivalent to the effect of internal H_{10} curve of the second case. The new part of this case is the influence of both H_{10} curves in Region e. There are four Hopf points along the S-shaped steady-state locus. Five different bifurcation diagrams are identified inside this region:

• e(1):   only the Hopf point at the longer residence time (number 4) is subcritical;

• e(2):   because of crossing both H_{10} curves at the point of the tangency, the first and the third Hopf points become subcritical;

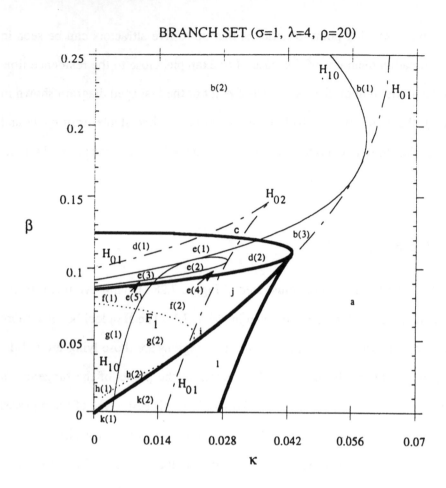

Figure 7.28. Branch set for $\sigma = 1$, $\lambda = 4$, $\rho = 20$.

- e(3): crossing only the first (external) $\mathbf{H_{10}}$ curve results in changing the stability of the limit cycles emerging from the first Hopf point;

- e(4): all Hopf points become subcritical in this region;

- e(5): only stable limit cycles emerges from the third Hopf point .

Case 4: $\sigma = 1$, $\lambda = 0.04$, $\rho = 5$

Similarly to the previous cases, we have two H_{10} curves. The effect of these H_{10} loci can be examined in a similar way to the previous cases. Combination of these curves with the other codimension-one singularities results in dividing the κ, β–plane into twenty regions. The branch set for this case is given in Figure 7.29 and the typical forms for these bifurcation diagrams (a-n) are shown in Table 7.1. In comparison with the previous cases, only two new patterns are found. These are mushroom-type k(2) and isola type m(2). In these regions, the steady state loci have only two supercritical Hopf points lying on the bottom part of the loci.

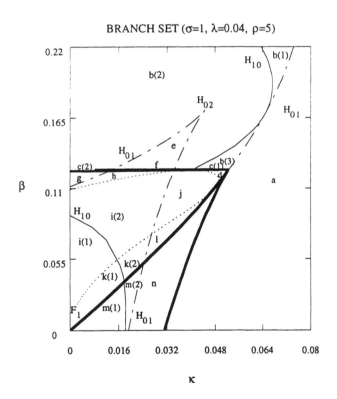

Figure 7.29. Branch set for $\sigma = 1$, $\lambda = 0.04$, $\rho = 5$.

165

Case 5: $\sigma = 1, \lambda = 400, \rho = 5$

Putting the H_{10} loci in the branch set for this case does not lead any new patterns. The κ–β–parameter region is divided into twelve regions (Figure 7.30). The typical forms for the bifurcation diagrams of these regions are given in Table 7.1.

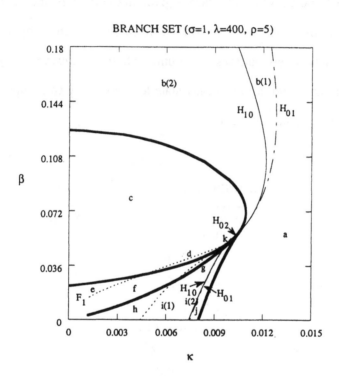

BRANCH SET ($\sigma=1$, $\lambda=400$, $\rho=5$)

Figure 7.30 Branch set for $\sigma = 1, \lambda = 400, \rho = 5$

Case 6: $\sigma = 0.5, \lambda = 9, \rho = 5$

Twenty five bifurcation diagrams have been confirmed for this case. The branch set is given in Figure 7.31. Two more new steady-state patterns are found. The first pattern is in Region e, where the model shows an isola-type equilibrium locus with two

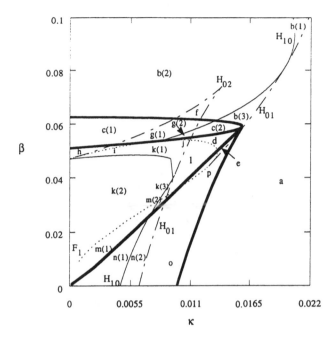

Figure 7.31 Branch set for $\sigma = 0.5$, $\lambda = 9$, $\rho = 5$

subcritical Hopf points lying in the top equilibrium curve. The second one corresponds to

Region j, the model exhibits four Hopf bifurcation points; the second and the third points

which lie in the bottom part of the steady-state locus are supercritical. See Table 7.1 for

the forms of the bifurcation diagrams for this case.

Case 7: $\sigma = 2$, $\lambda = 2.25$, $\rho = 5$

The branch set for this case is shown in Figure 7.32. No more new bifurcation

diagrams has been found for this case. Table 7.1 gives the forms of the steady-state loci

which correspond to the twelve regions found for this case.

167

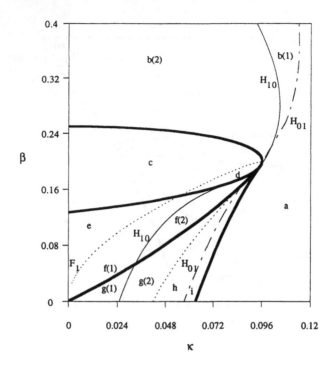

Figure 7.32. Branch set for $\sigma = 2.0$, $\lambda = 2.25$, $\rho = 5$.

We have seen that two H_{10} curves are specified in each case. Table 7.1 shows that forty one bifurcation diagrams have been confirmed for this model (so far). It should be read in conjunction with Figure 7.33.

7.4.2 H_{20}

This type of degeneracy leads to two limit points on the periodic branch and three periodic solutions. The unfoldings and normal forms of H_{20} degeneracies are discussed in Section 6.3.2. The formula of Golubitsky and Langford [1981] are used for studying these singularities in the parameter space. H_{20} is of codimension two, which means that

168

Bifurcation Type (see Figure 7.33)	Case 1	Case 2	Case 3	Case 4	Case 5	Case 6	Case 7
1	a	a	a	a	a	a	a
2	e(1)	b(1)	b(1)	b(1)	b(1)	b(1)	b(1)
3	e(2)	b(2)	b(2)	b(2)	b(2)	b(2)	b(2)
4		b(3)	b(3)	b(3)		b(3)	
5			c	e		f	
6	b						
7	f						
8	g	c(1)	d(1)	c(2)	c	c(1)	c
9		c(2)	d(2)	c(1)		c(2)	
10	h				d		d
11		d	e(1)	f		g(1)	
12			e(2)				
13			e(3)			g(2)	
14			e(4)				
15			e(5)				
16	c						
17		j	j	l			
18	i	f		g	e	h	e
19	j(1)				f		f(1)
20	j(2)						f(2)
21		e		h		i	
22		g(1)	g(1)	i(1)		k(2)	
23		g(2)		i(2)		k(1)	
24		g(3)	g(2)			k(3)	
25				k(2)	g		
26		h(1)	h(1)	k(1)		m(1)	
27		h(2)	h(2)			m(2)	
28		i	i	j		l	
29		m		d		d	
30			f(1)				
31			f(2)				
32						j	
33	d						
34	k(1)				h	p	g(1)
35	k(2)						g(2)
36					i(1)		
37		k(1)	k(1)	m(1)		n(1)	
38		k(2)	k(2)		i(2)	n(2)	h
39				m(2)			
40		l	l	n	j	o	i
41						e	

Table 7.1. Bifurcation diagrams for the seven cases.

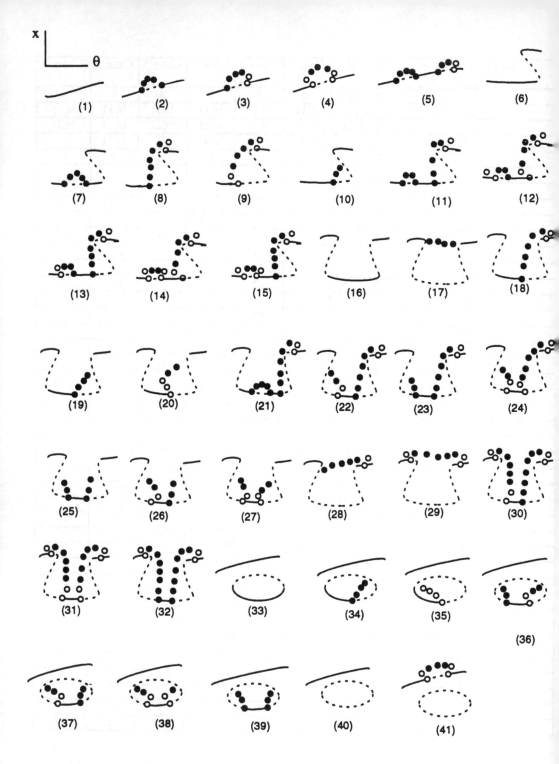

Figure 7.33. Bifurcation diagrams for the first model.

170

it will be represented as a point in the codimension-one singularities diagram. Around this point the plane is divided by three transition varieties, namely H_{01}, a double limit point variety (limit point lined up with Hopf point) and a hysteresis variety (see Figure 6.4).

We have examined the H_{20} points for the seven cases studied before. No H_{20} points are found on the upper H_{10} curve. This means that no unique steady-state bifurcation diagrams with three simultaneous periodic solutions exists for these seven cases. We have found one H_{20} point sitting on the lower H_{10} curve (which crosses the multiplicity regions) in four of the seven cases (Figure 7.34). They are case 1 ($\sigma = 1$, $\lambda = 4$, $\rho = 1$), case 2 ($\sigma = 1$, $\lambda = 4$, $\rho = 5$), case 3 ($\sigma = 1$, $\lambda = 4$, $\rho = 20$) and case 7 ($\sigma = 2$, $\lambda = 2.25$, $\rho = 5$). In all these cases, this degeneracy takes place at the larger residence time Hopf point of the bottom part of the mushroom or isola diagrams. Close to these points, the model is expected to show multiple periodic solutions. As an example, two stable limit cycles are obtained for the second case for conditions close to the H_{20} point (Figure 7.35). In Figure 7.36, a new curve, hysteresis variety, starting from the H_{20} point divide the Regions f(1), g(1), and h(1) of the branch set of the third case. Three periodic solutions can be found for parameter values lying to the right of this curve and to the left of the H_{10} curve. The complete unfoldings of the H_{20} singularity is shown in Figure 7.37 for the seventh case. The double-limit point and the hysteresis varieties are added the branch set of the case ($\sigma = 2$, $\lambda = 2.25$, $\rho = 5$). Figure 7.38 shows the bifurcation diagrams computed for $\beta = 0.0667$ with κ traversing all four regions of the H_{20} unfolding. The stable periodic solutions for the case, $\beta = 0.0667$, $\kappa = 0.0022$, are plotted in Figure 7.39.

Some difficulties have been encountered in finding multiple periodic solutions around the H_{20} points for the first and the second cases which can be explained by the

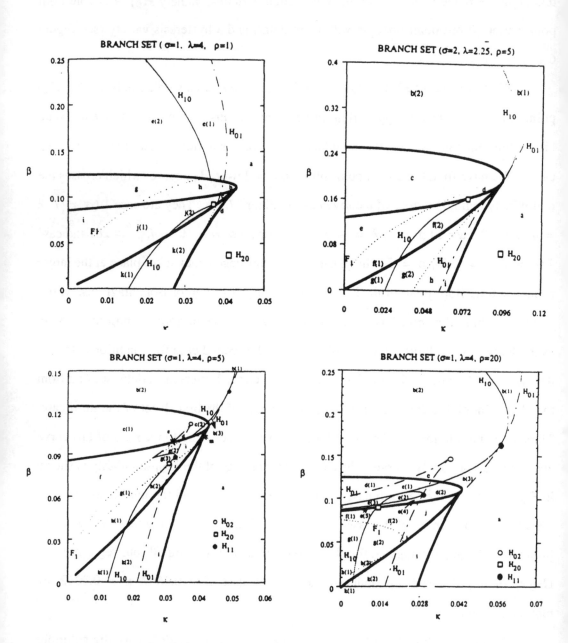

Figure 7.34. **H₂₀** singularities in the branch sets.

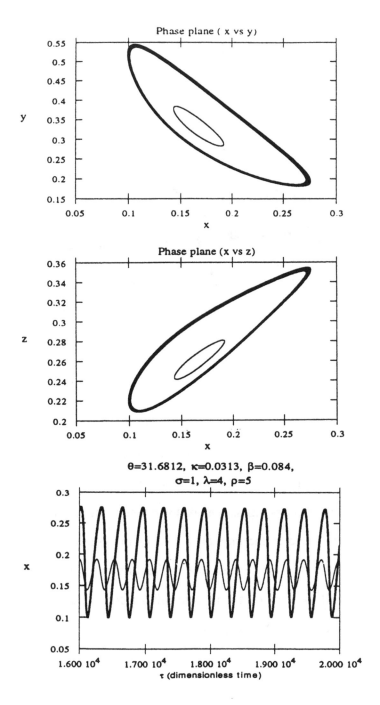

Figure 7.35. Stable limit cycles.

173

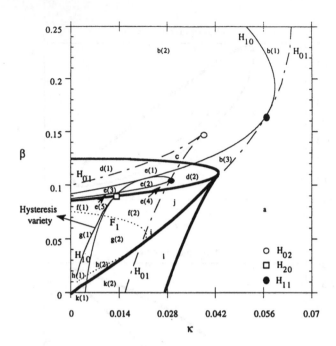

Figure 7.36. Branch set for case 3.

following. This can be explained by noticing that close to the isola-mushroom boundaries, there are some interactions between the middle steady-state curve (saddle-node) and the periodic solutions emerging from the Hopf points. The periodic solutions collide with these saddle-node points and break down before forming the hysteresis-shape periodic branches. Because of the implicit nature of the conditions of the third type degeneracies, we were not able to compute entire branch of this codimension-two singularity. This will be very useful in determining for which σ, λ, ρ values the codimension-one branch set has H_{20} singularities.

174

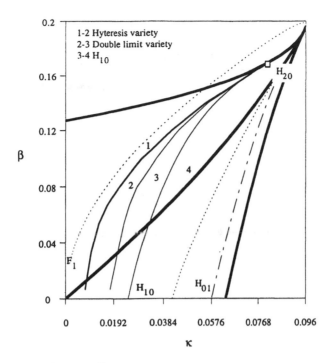

Figure 7.37. Branch set for case 7.

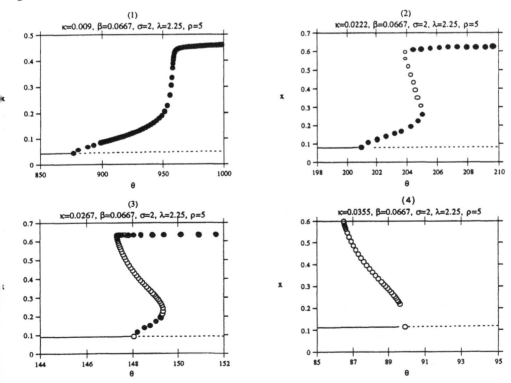

Figure 7.38. Bifurcation diagrams around the H_{20} point - case 7.

175

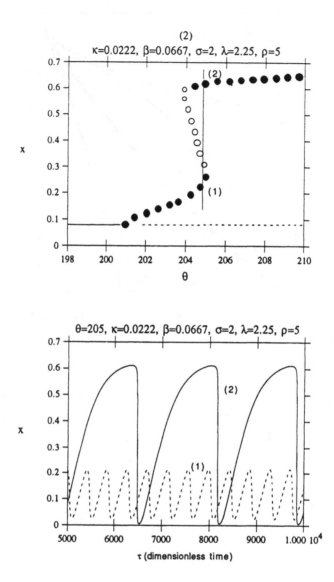

Figure 7.39. Stable periodic solutions around the H_{20} point - case 7.

7.4.3 H_{11}

H_{11} is one of the most interesting degeneracies obtained for this model. It occurs when the H_{10} curve intersects the H_{01} curve tangentially. Two forms of H_{11} degeneracies have

176

been observed in the seven cases. The first one is equivalent to the structure of the H_{11} point corresponding to the case $\varepsilon = 1$, $\gamma = -1$, $m < 0$ (see Figure 6.5). This kind is found around the point of tangency of the internal H_{10} and H_{01} curves in cases 2 and 3 (see Figure 7.40). Consider the two Hopf points lying on the bottom steady state solution curve of the regions $g(2)$, $g(1)$, $g(3)$ and i in the second case shown in Figure 7.27. Comparing these regions with the bifurcation diagrams in Figure 6.5, we notice

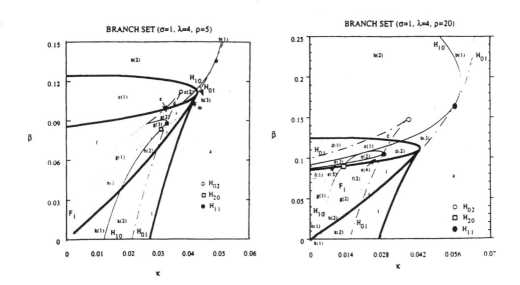

Figure 7.40. Branch sets for case 2 and case 3.

that another locus corresponding to the β_∞ varieties can be found. Here the periodic branches emerging from both Hopf points passes through a point where an isolated periodic solution branch is formed. This point is called transcritical bifurcation. In Figure 7.41, the steady-state locus diagram shows an isolated periodic branch formed by

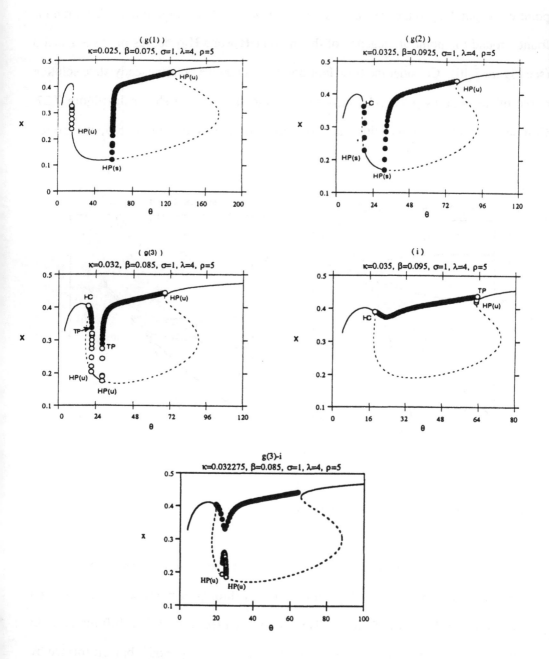

Figure 7.41. Isolated periodic solution - transcritical bifurcation.

178

simple or transcritical bifurcation. Similar conclusion can be drawn for the case ($\sigma = 1$, $\lambda = 4$, $\rho = 20$) using the regions $d(2)$, $e(3)$, $e(2)$ and $e(4)$.

The second H_{11} occurs at the unique region where the upper H_{10} and H_{11} curves move very close to each other. This kind of H_{11} is found for the cases: case 2 ($\sigma = 1$, $\lambda = 4$, $\rho = 5$), case 3 ($\sigma = 1$, $\lambda = 4$, $\rho = 20$), case 4 ($\sigma = 1$, $\lambda = 0.04$, $\rho = 5$), and case 6 ($\sigma = 0.5$, $\lambda = 2.25$, $\rho = 5$) (Figure 7.42). These tangential intersections agree with the unfoldings constructed by Golubitsky and Langford using singularity theory. The normal form and the unfolding for this case is summarized in Figure 6.7. There exits a third curve emerging from the H_{11} which corresponds to the birth of the isolas of periodic solutions. On the branch set of our model, this locus lies below the H_{01} structure, extending from the H_{11} point towards smaller β values. This curve is shown in the branch set of the third case in Figure 7.43. Isola of periodic solutions are shown for the case $\beta = 0.15$, $k = 0.05383$ in Figure 7.44.

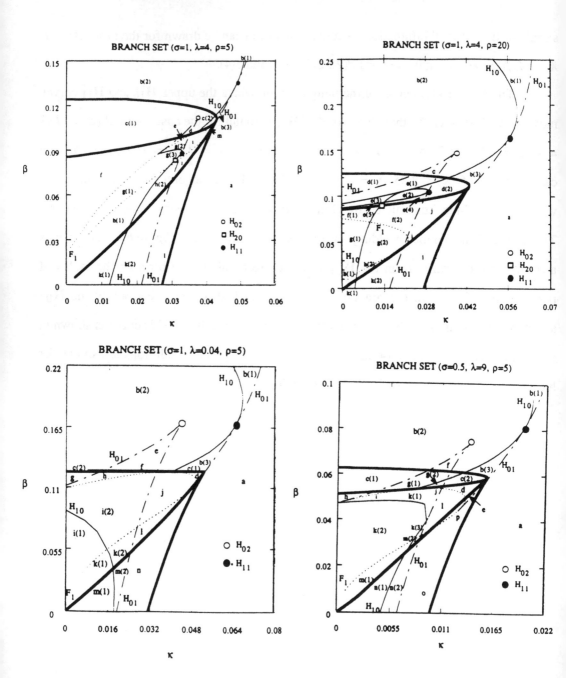

Figure 7.42. Branch sets for cases 2, 3, 4, and 6.

180

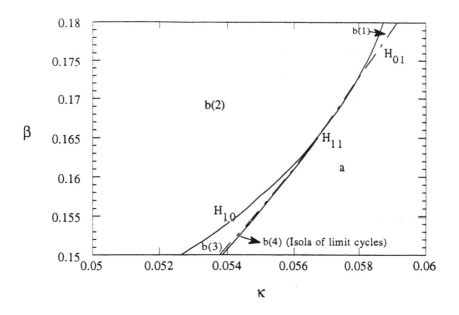

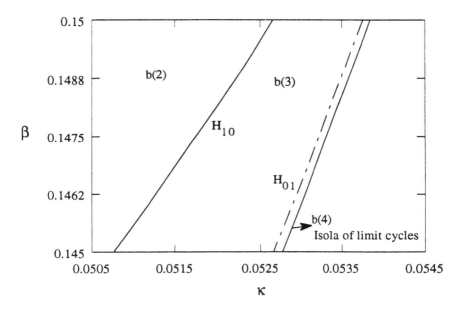

Figure 7.43. Branch set for case 3.

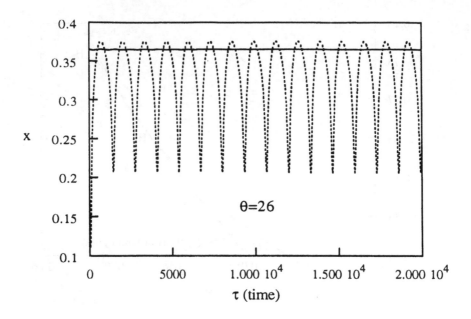

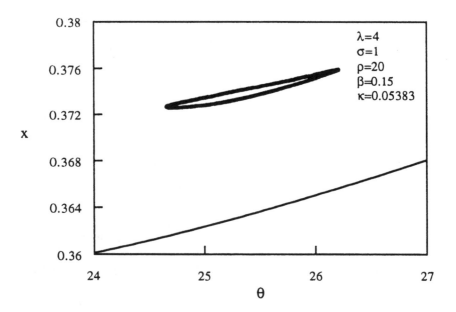

Figure 7.44. Bifurcation diagram - isola of periodic solutions.

182

Chapter 8

DYNAMICS - MODEL II

The dynamic behaviour of the second model (coupling through the autocatalyst) is investigated in this chapter. In the first section we give the Hopf bifurcation conditions in terms of the characteristic equation's coefficients. The second section of this chapter deals with the codimension one singularities. They are used to subdivide the parameter space into regions where samples of different bifurcation diagrams can be found. The third section examines the F_2 degeneracies using the normal form method to define the solutions around these degeneracies. Finally, we present some of the codimension-two degeneracies which are obtained for this model.

8.1   Hopf Bifurcation

The mathematical model, in a dimensionless form, is:

$$\frac{dx}{d\tau} = \frac{1 - x}{\theta} - xy^2$$

$$\frac{dy}{d\tau} = \frac{z - y}{\lambda} - \frac{y}{\theta} + xy^2 - \kappa y$$

$$\rho \frac{dz}{d\tau} = -\frac{z - y}{\lambda} + \frac{\sigma}{\theta} (\beta - z) - \rho \kappa z \tag{8.1}$$

The Jacobian matrix J of this model is:

$$J = \begin{pmatrix} -\dfrac{1}{\theta} - y_{ss}^2 & -2x_{ss}y_{ss} & 0 \\[2ex] y_{ss}^2 & -\dfrac{1}{\lambda} - \dfrac{1}{\theta} + 2x_{ss}y_{ss} - \kappa & \dfrac{1}{\lambda} \\[2ex] 0 & \dfrac{1}{\rho\lambda} & -\dfrac{1}{\rho\lambda} - \dfrac{\sigma}{\rho\theta} - \kappa \end{pmatrix} \qquad (8.2)$$

The eigenvalues μ of the Jacobian matrix are the solutions of the characteristic equation:

$$-\mu^3 + S_1\mu^2 - S_2\mu + S_3 = 0 \qquad (8.3)$$

where S_1, S_2, and S_3 are the three invariants of J

$$S_1 = j_{11} + j_{22} + j_{33} \qquad (8.4)$$

$$S_2 = \det\begin{pmatrix} j_{11} & j_{12} \\ j_{21} & j_{22} \end{pmatrix} + \det\begin{pmatrix} j_{22} & j_{23} \\ j_{32} & j_{33} \end{pmatrix} + \det\begin{pmatrix} j_{11} & j_{13} \\ j_{31} & j_{33} \end{pmatrix} \qquad (8.5)$$

$$S_3 = \det(J) \qquad (8.6)$$

where the terms $j_{11}, j_{12}, \ldots$, are the elements of J. The general conditions for Hopf bifurcation for three-dimensional models are:

$$F = S_1S_2 - S_3 = 0 \qquad (8.7)$$

$$\omega^2 = S_2 > 0 \qquad (8.8)$$

184

The condition F is found to be quartic in x_{ss}, which means up to four Hopf points can be found in this model. These four possible Hopf points can be found in various configurations in the bifurcation diagrams. One of these forms is shown in Figure 8.1, where the model has unique steady state for all θ values. As a result of the appearance of four Hopf points in the bifurcation diagram, the steady state locus is subdivided into five segments. For this case, the model exhibits stable, unstable steady states and periodic solutions.

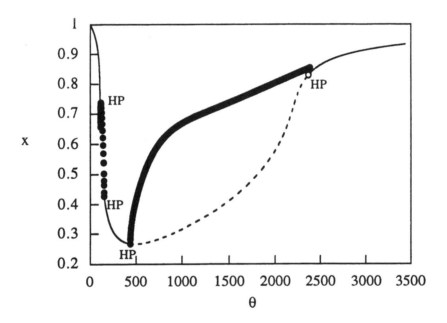

Figure 8.1. Bifurcation diagram for the case $\kappa = 0.006$, $\beta = 0.03$, $\sigma = 5.0$, $\lambda = 1.0$, $\rho = 1.0$.

8.2 Codimension-one Singularities

We begin analyzing this model by considering the following codimension-one degeneracies:

• Static bifurcations: they define the conditions of the appearance/disappearance of multiple steady states in the bifurcation diagrams. The static singularities take care of the turning point-turning point interactions. The hysteresis and isola-mushroom singularities are the only static singularities obtained for this model.

• Double-zero eigenvalues F_1: this results from the simplest interactions of Hopf point and a limit point. The number of Hopf points changes by one when passing through this kind of degeneracy. The Jordan block structure for this case is

$$\begin{pmatrix} 0 & 1 \\ 0 & 0 \end{pmatrix}$$

It can be shown easily that the second model has only this kind of Jordan block by constructing a 2 x 2 matrix that always has nonzero determinant. The conditions for F_1 are:

$$S_2 = S_3 = 0 \tag{8.9}$$

• H_{01}: this corresponds to the appearance or coalescence of two Hopf points in the bifurcation diagrams. The conditions for H_{01} are

$$F = F_x = 0, \quad F_{xx} \neq 0 \tag{8.10}$$

where $F = S_1 S_2 - S_3$, $S_2 > 0$.

H$_{10}$: this describes the condition for the change in the stability of the emerging periodic solutions at a Hopf point. The formulae of Golubitsky and Langford [1981] are used to follow this type of degeneracy.

We examine these singularities for nine cases. The resulted bifurcation patterns are summarized in Table 8.1 (Table 8.1 will be found at the end of this section.)

1. $\sigma = 1.0,\ \lambda = 1.0,\ \rho = 1.0$

    Tho oodimension-one singularities for this case are shown in Figure 8.2. The branch set for this case is qualitatively equivalent to the case of single CSTR. Fourteen

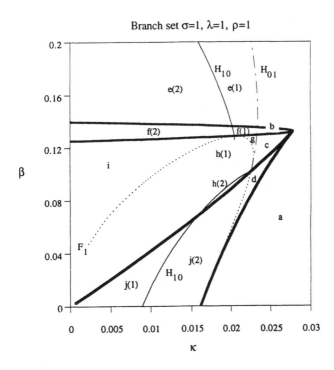

Figure 8.2. Branch set for the case $\sigma = 1.0,\ \lambda = 1.0,\ \rho = 1.0$.

bifurcation diagrams are distinguished for this case (Figure 8.3). First, for large values of κ the model exhibits stable unique steady state for all β (Region a). While the regions b, c, d, have similar behaviour to each other, with one qualitative difference: no oscillatory behaviour is found, but there is a multiplicity of steady states. In (b) we have a hysteresis; in (c) the steady state locus forms a mushroom and in (d) we have an isola. $\mathbf{H_{01}}$ curve starts at a point of tangency between the $\mathbf{F_1}$ curve and the isola-mushroom boundary. Crossing this curve from right to left (i.e., using smaller κ values) results in the appearance of two supercritical Hopf points in the bifurcation diagrams of the unique, hysteresis, and mushroom patterns (e(1), f(1) and g). When using lower κ, the limit cycles emerging from the Hopf points at the larger θ in e(1) and f(1) change their stability as we pass through the $\mathbf{H_{10}}$ curve and this results in the bifurcation diagrams e(2) and f(2). The periodic solution branch in these regions undergo the simplest secondary dynamic bifurcation, saddle-node bifurcation or turning point (TP). To the left of this (TP) point, the model exhibits stable and unstable periodic solutions. In Region (i), the mushroom pattern has similar dynamic behaviour to Region f(2). The Hopf point at the larger θ in Region (g) disappears when crossing the $\mathbf{F_1}$ curve to Region h(1). In h(1), the periodic solutions emanating from the remaining Hopf point die at saddle-node points (Homoclinic orbit, HC). The isola pattern in j(1) exhibits similar dynamic behaviour to Region h(1). Figure 8.2 shows another $\mathbf{H_{10}}$ curve in the bottom part. It passes through the isola and mushroom regions and ends at the $\mathbf{F_1}$ curve. As a result of crossing this curve, the steady state patterns in Regions h(2) and j(2) show subcritical Hopf points. Clearly, one can see that no regions with more than two Hopf points are found for this case.

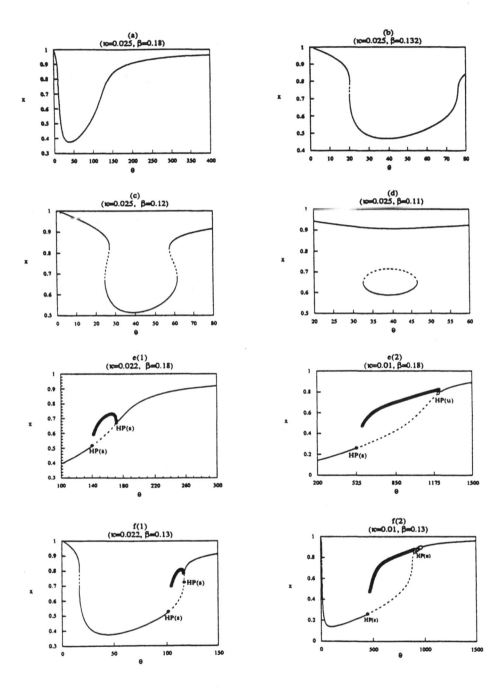

Figure 8.3. Bifurcation diagrams for the case $\sigma = 1.0$, $\lambda = 1.0$, $\rho = 1.0$.

Figure 8.3 (continued)

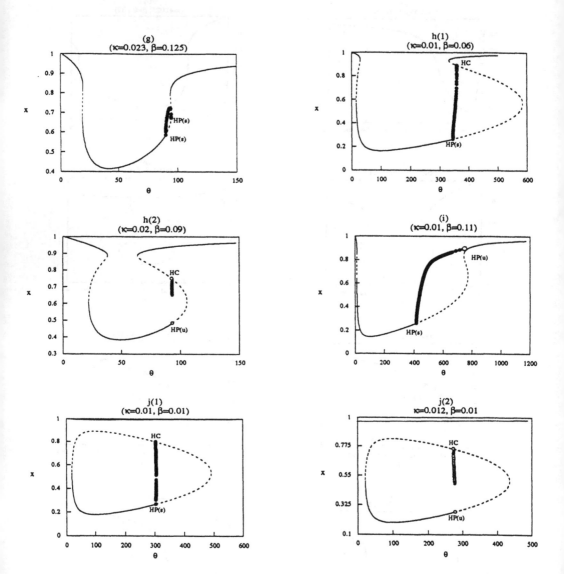

$\sigma = 0.5, \lambda = 1.0, \rho = 1.0$

Decreasing the parameter σ to 0.5 does not add any new bifurcation diagram for this model. Figure 8.4 shows the branch set for this case. It can be seen the model exhibits multiple equilibrium solutions for larger κ and β values in comparison with the first case. Both cases start exhibiting dynamic behaviour at similar values of κ and β.

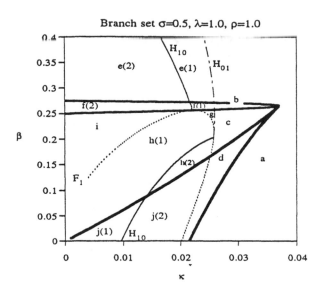

Figure 8.4. Branch set for the case $\sigma = 0.5, \lambda = 1.0, \rho = 1.0$

3. $\sigma = 5.0, \lambda = 1.0, \rho = 1.0$

Figure 8.5 shows that using higher σ has resulted in an increase of the number of the bifurcation diagrams to twenty two. Seventeen of these configurations are new for this model (Figure 8.6). A second H_{01} curve has appeared on the left side of the branch set. It has a sharp turning point which is denoted by H_{02}. In Regions c, g and j four Hopf points appear in the bifurcation diagrams. Three of these Hopf points are supercritical while the Hopf point which lies at the larger residence time is

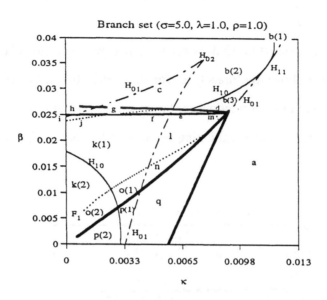

Figure 8.5. Branch set for the case $\sigma = 5.0$, $\lambda = 1.0$, $\rho = 1.0$

subcritical. It can be seen that the upper H_{10} curve touches the H_{01} curve at a point where a higher order degeneracy is found, namely H_{11}. This curve H_{10} ends at a point of tangency between the F_1 and the hysteresis-unique boundary. Around the H_{11} point, the unique steady-state patterns exhibit three different dynamic behaviour; b(1), b(2) and b(3). Throughout Region d, the model exhibits a new hysteresis pattern with two subcritical Hopf points. If we decrease κ we pass through the right side of F_1 degeneracies which leads to Region e. In this region, the behaviour is slightly different from that in Region d, showing a hysteresis behaviour with only one Hopf point. Again, decreasing κ transgresses the H_{01} curve and results in the appearance of two more supercritical Hopf points, as shown in Region f. If we reduce the parameter κ further, we cross the left side of F_1 and a fourth Hopf point appears close to the lower turning point, (see Region g). We reach Region h by

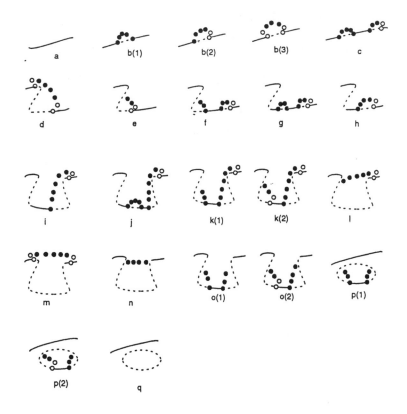

Figure 8.6. Bifurcation diagrams for the case $\sigma = 5.0$, $\lambda = 1.0$, $\rho = 1.0$

using lower decay rate parameter. Here, two Hopf points disappear when crossing the other side of the H_{01} curve. In the Regions m, l, k(1), j and i the steady state loci show mushroom patterns which have similar dynamic behaviour to the Regions d, e, f, g and h, respectively. One of the most interesting bifurcation diagrams is found through the Region n. In this case, there are no Hopf points but there are periodic solution branches for θ values at which all the equilibrium points are unstable. Here the limit cycles in both sides of the branch die at saddle-node points. Throughout the Region o(1), the model has a mushroom with two supercritical Hopf points lying on the bottom part of the steady state locus. In Regions k(2) and o(2), the model exhibits a mushroom pattern similar to k(1) and o(1), respectively, with only one difference. Unstable limit cycles emerge from the Hopf points which are locating at the smallest residence time. In Region q, the model has an isola pattern but it does not show any dynamic motions. In the final two Regions, p(1) and p(2), the steady state loci show an isola shape with similar dynamic behaviour to the forms found in Regions o(1) and o(2), respectively.

4. $\sigma = 5.0, \lambda = 1.0, \rho = 0.1$

The κ, β–plane is divided into seventeen regions as shown in Figure 8.7. The model has similar behaviour to the third case, with one difference. There is only one H_{10} curve in the branch set. Three more new bifurcation forms can be identified. In

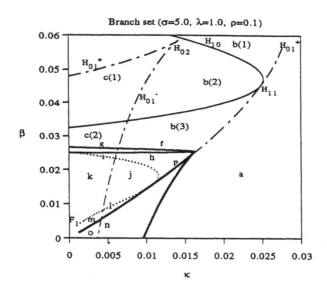

Figure 8.7. Branch set for the case $\sigma = 5.0, \lambda = 1.0, \rho = 0.1$

regions c(2) and *g* we have unique and hysteresis forms with four Hopf points. The Hopf points which lie at the smallest and the largest residence times are subcritical. In Region p, the model has an isola-type equilibrium form with two subcritical Hopf points lying on the top equilibrium curve. A comparison between this case and case three shows that models with lower ρ values exhibit static and dynamic bifurcations at larger κ and β values.

5. $\sigma = 5.0, \lambda = 1.0, \rho = 10.0$

The branch set shown in Figure 8.8 is similar to the parameter space of the first case. No new bifurcation diagram is found for this case.

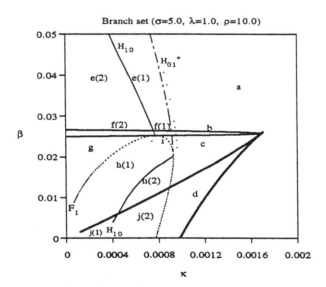

Figure 8.8. Branch set for the case $\sigma = 5.0, \lambda = 1.0, \rho = 10.0$

6. $\sigma = 5.0, \lambda = 0.1, \rho = 1.0$

The loci of degenerate bifurcation in the κ, β–plane (Figure 8.9) are qualitatively similar to those of the third case. Therefore, reducing the mass transfer resistance parameter from 1 to 0.1 does not lead to any new pattern.

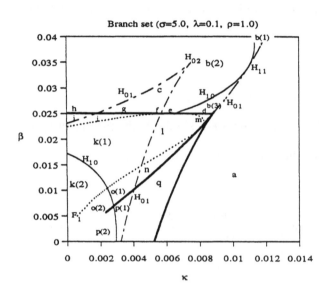

Figure 8.9. Branch set for the case $\sigma = 5.0, \lambda = 0.1, \rho = 1.0$

7. $\sigma = 5.0, \lambda = 10.0, \rho = 1.0$

The branch set for this case is illustrated in Figure 8.10. Twenty five bifurcation diagrams have been confirmed (Figure 8.11). Six new forms are identified. In Region c, the model has a unique steady state with four supercritical Hopf points. In Regions f, g(1), h(1), i(1), and i(2) we have hysteresis forms with new dynamic behaviour. In contrast to case 3, we have found that increasing λ enlarged the area in the κ, β-plane which is affected by static or dynamic degeneracies.

196

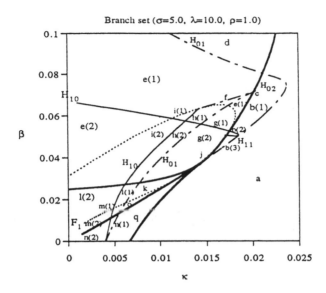

Figure 8.10. Branch set for the case $\sigma = 5.0$, $\lambda = 10.0$, $\rho = 1.0$

Next we examine the dynamic degeneracies of the model around a point where the hysteresis curve has more than one turning point. In Chapter 5, we have found that at $\sigma = 8$ and $\lambda > 8$ the hysteresis curve has three turning point in the κ, β-plane. In the following two cases we study the model around this point.

8. $\sigma = 8.0$, $\lambda = 7.0$, $\rho = 1.0$

The branch set for this case (Figure 8.12) is similar to the forth case. Eighteen bifurcation diagrams are identified but no new patterns are found.

197

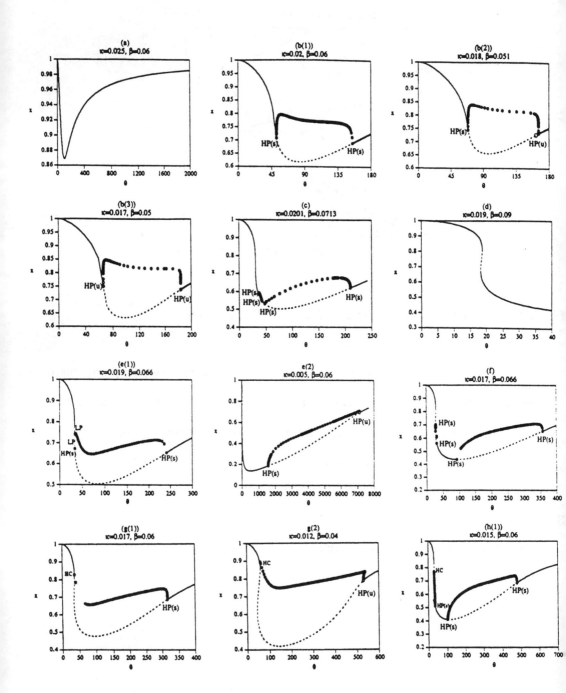

Figure 8.11. Bifurcation diagrams for the case $\sigma = 5.0$, $\lambda = 10.0$, $\rho = 1.0$

Figure 8.11 (continued)

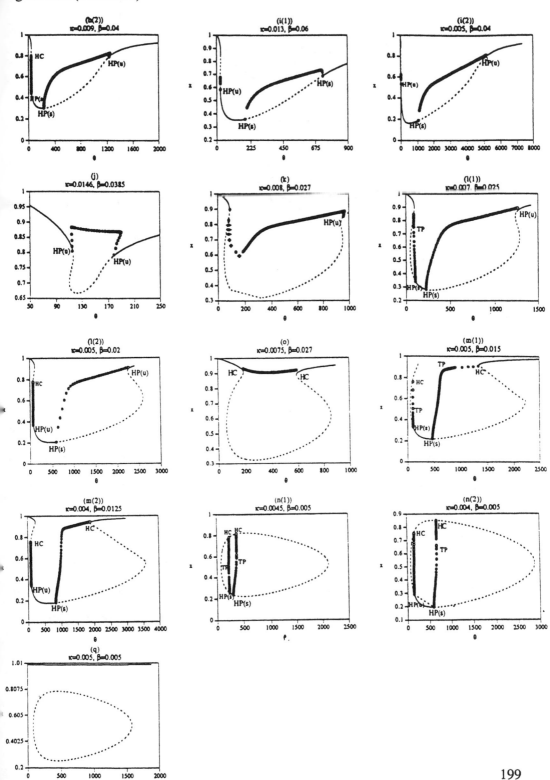

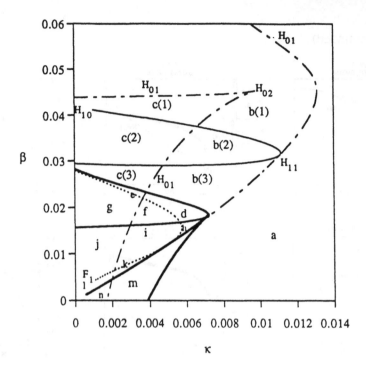

Figure 8.12. Branch set for the case $\sigma = 8.0$, $\lambda = 7.0$, $\rho = 1.0$

9. $\sigma = 8.0$, $\lambda = 9.0$, $\rho = 1.0$

Figure 8.13 shows that the κ, β–plane is divided into twenty three regions. The change in the hysteresis curve has resulted in five patterns not found in the previous case, though known from elsewhere. Note also that the dynamic degeneracies have similar forms to the previous case ($\lambda = 7.0$).

Finally, we summarize all the bifurcation forms obtained for this model in Table 8.1 and Figure 8.14. Forty bifurcation diagrams have been confirmed for this model. More bifurcation diagrams can be found by studying codimension two-singularities.

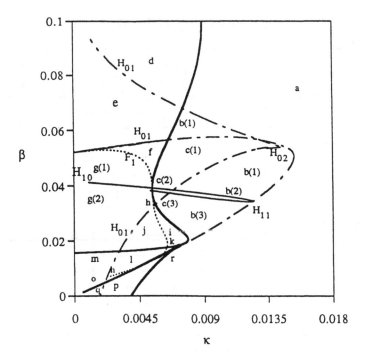

Figure 8.13. Branch set for the case $\sigma = 8.0$, $\lambda = 9.0$, $\rho = 1.0$.

Bifurcation Type (see Figure 8.14)	Case 1,2	Case 3,6	Case 4	Case 5	Case 7	Case 8	Case 9
1	a	a	a	a	a	a	a
2	e(1)	b(1)	b(1)	e(1)	b(1)	b(1)	b(1)
3	e(2)	b(2)	b(2)	e(2)	b(2)	b(2)	b(2)
4		b(3)	b(3)		b(3)	b(3)	b(3)
5		c	c(1)			c(2)	c(2)
6					c	c(1)	c(1)
7			c(2)			c(3)	c(3)
8	b			b	d		d
9	f(1)			f(1)	e(1)		e
10	f(2)	h		f(2)	e(2)		f
11		d	f		j	d	i
12		e			g(2)	f	j
13		f			h(2)	g	g(2)
14		g					
15					f		
16					g(1)		
17					h(1)		g(1)
18					i(1)		
19					i(2)		
20			g			e	h
21	c			c			
22	g			i			
23	h(1)			h(1)			r
24	h(2)			h(2)			
25	i	i		g			
26		j					
27		k(1)	k		l(1)		m
28		k(2)			l(2)	j	
29		l	j		k	i	l
30		m	h			h	k
31		n	l		o	k	n
32		o(1)	i		m(1)	l	o
33		o(2)	m		m(2)		
34	d			d			
35	j(1)			j(1)			
36	j(2)			j(2)			
37		p(1)	o		n(1)	n	q
38		p(2)			n(2)		
39		q	n		q	m	p
40			p				

Table 8.1. Bifurcation diagrams for Model II.

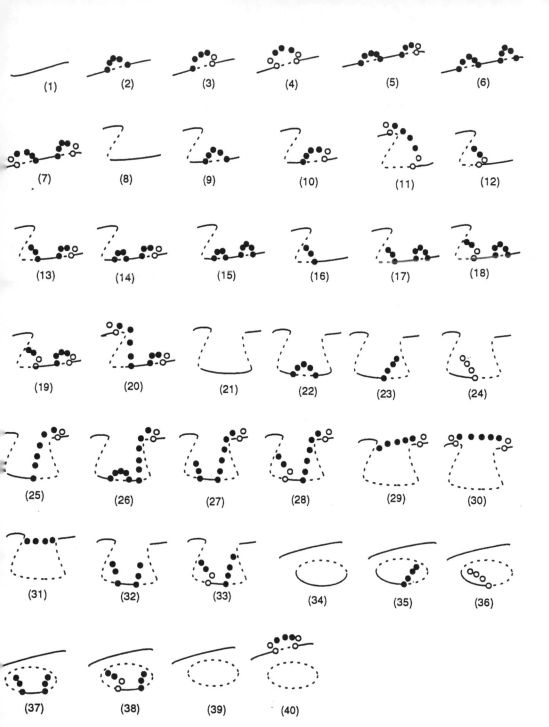

Figure 8.14. Bifurcation patterns for Model II.

203

8.3   F₂: Pure Imaginary Pair and Simple Zero Eigenvalue

The $\mathbf{F_2}$ degeneracies occur when the linearized model has a simple bifurcation of steady states (such as turning point) and a Hopf bifurcation existing at the same point in the parameter space.   The conditions for $\mathbf{F_2}$ degeneracies in terms of the coefficients of the characteristic equation are defined as

$$S_1 = S_3 = 0, \quad S_2 = \omega^2 > 0 \tag{8.11}$$

These coefficients for the second model are given by

$$S_1 = j_{11} + j_{22} + j_{33} \tag{8.12}$$

$$S_2 = j_{11}j_{22} + 2x_{ss}y_{ss}^3 + j_{22}j_{33} - \frac{1}{\rho\lambda^2} + j_{11}j_{33} \tag{8.13}$$

$$S_3 = j_{11}\left(j_{22}j_{33} - \frac{1}{\rho\lambda^2}\right) + 2xy^3j_{33} \tag{8.14}$$

First we show that $\mathbf{F_2}$ bifurcation occurs for this model.   Assume that the equality conditions hold for some parameter values, and check the sign of the coefficient S_2 under these assumptions. Since $j_{33} < 0$ for positive parameters values, the condition $S_2 > 0$ is equivalent to the condition

$$j_{33}S_2 = j_{11}j_{22}j_{33} + 2x_{ss}y_{ss}^3j_{33} + j_{22}j_{33}^2 - \frac{1}{\rho\lambda^2}j_{33} + j_{11}j_{33}^2 < 0$$

We use the condition equation 8.14 to substitute for the first two terms of this condition. The inequality above can be written as

$$j_{33}S_2 = \frac{1}{\rho\lambda^2} (j_{11} - j_{33}) + j_{33}^3 (j_{11} + j_{22}) < 0$$

Then, we use the condition $S_1 = 0$ (equation 8.12), to obtain the simplified expression

$$j_{33}S_2 = \frac{1}{\rho\lambda^2} (j_{11}j_{33}) - j_{33}^3 < 0$$

or

$$\frac{1}{\rho\lambda^2} (j_{11} - j_{33}) < j_{33}^3$$

This shows that the condition $S_2 > 0$ depends on the sign of j_{11} and j_{33}. Both quantities have negative values for positive parameters which means that it is possible to find $\mathbf{F_2}$ degeneracies for the second model.

The loci of the $\mathbf{F_2}$ bifurcations in the parameter space can be found by the solving the conditions equation 8.11, and the steady state equations simultaneously. The $\mathbf{F_2}$ in the branch set of the case $\sigma = 0.1$, $\lambda = 20.0$, $\rho = 20.0$ in Figure 8.15 are obtained by varying the decay rate constant κ and solving the $\mathbf{F_2}$ conditions for x, θ, and β. The inequality condition for $\mathbf{F_2}$ bifurcation is violated at two points in this branch diagram. At these points, $\mathbf{F_2}$ and $\mathbf{F_1}$ intersect which results in the appearance of a codimension-two, triple-zero eigenvalues or $\mathbf{G_1}$ degeneracies. Actually, the occurrence of $\mathbf{G_1}$ bifurcation can be looked upon as a natural consequence of $\mathbf{F_2}$ and $\mathbf{F_1}$ intersections. Figure 8.16 shows the loci of the $\mathbf{G_1}$ bifurcations in the ρ, κ– and ρ, β–planes for different values of λ and σ. It can be seen that increasing σ or λ reduces the minimum value of ρ at which the $\mathbf{G_1}$ bifurcation occurs. For ρ larger than this minimum value,

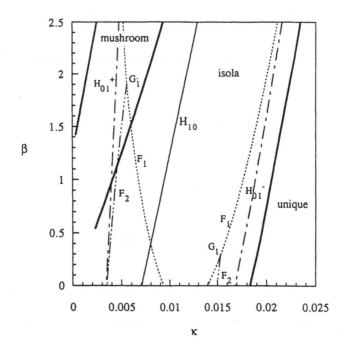

Figure 8.15. Branch set for the case $\sigma = 0.1$, $\lambda = 20.0$, $\rho = 20.0$.

the branch sets have two G_1 points. Perturbing the model around these singularities results in F_1 and F_2 degeneracies. On the other hand, using values of ρ smaller than the minimum leads to branch sets without G_1 or F_2 degeneracies.

Now we study the model behaviour around F_2 degeneracies in Figure 8.15. Two Hopf points appear in the bifurcation diagrams obtained near F_2 curves. Consider the F_2 curve which lies between the H_{01} and the left G_1 point. Close to these F_2 degeneracies, the bifurcation diagrams show two supercritical Hopf points. At the F_2 points, one Hopf points moves around the turning point which results in the bifurcation diagram shown in Figure 8.17. Torus bifurcation is found close to the F_2 point. Figure 8.18 shows that the model has quasiperiodic trajectories close to the torus bifurcation. While for the other

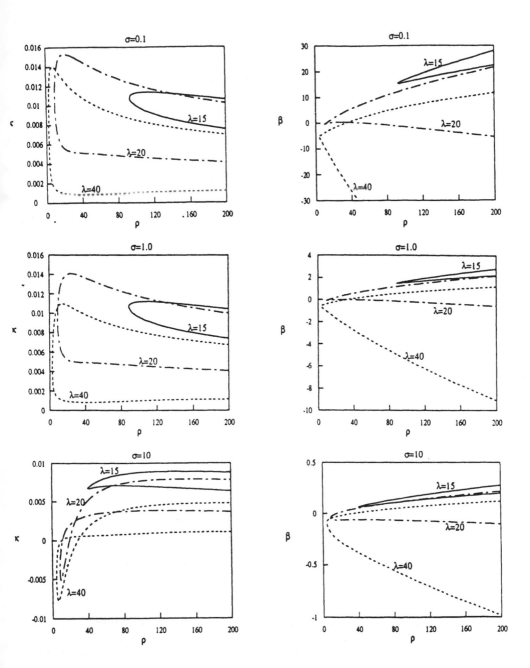

Figure 8.16. **G₁** singularities.

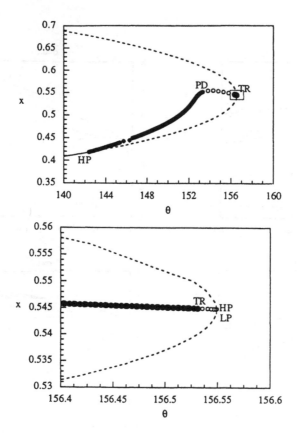

Figure 8.17. Bifurcation diagrams (κ = 0.00445, β = 1.0, σ = 0.1, λ = 20.0, ρ = 20.0).

F_2 curve which lies on the right of the branch set, the model has two subcritical Hopf bifurcation. In this case, the subcritical Hopf-saddle-node interactions does not lead to torus bifurcations. In fact, these types of behaviour have been found for the nonisothermal case of a first-order exothermic reaction in a CSTR with extraneous thermal capacitance [Planeaux and Jensen, 1986]. We now consider the periodic solutions emanating from the other Hopf points which lies at the smaller residence time.

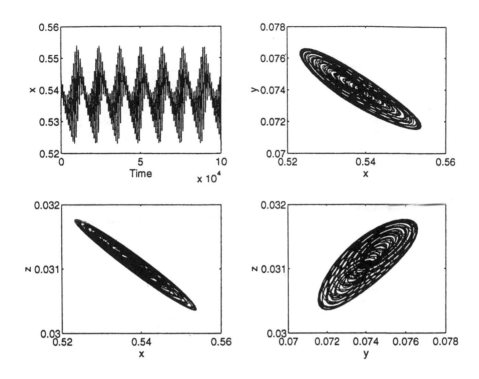

Figure 8.18. Times series and phase planes for (θ = 156.5220, κ = 0.00445, β = 1.0, σ = 0.1, λ = 20.0, ρ = 20.0).

Figure 8.19 a, b, c show the bifurcation diagrams for three cases along the left F_2 curve. It shows as β increases or as we move up toward the left G_1 point, the periodic solutions undergo period-doubling bifurcation. This means that the model can undergo higher period doubling which might leads to more complicated and subtle dynamic behaviour. This agrees with the fact that chaos arises in the G_1 vicinity. It can be also seen that period-doubling bifurcation occurs near the right G_1 point (see Figure 8.19d).

It remains now to use the normal form method to compute the torus branches and to examine their stability. The normal form transformation for the F_2 bifurcation and the

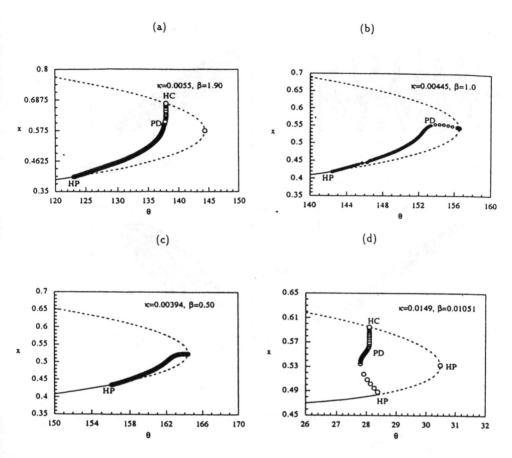

Figure 8.19. Bifurcation diagram $\sigma = 0.1, \lambda = 20.0, \rho = 20.0$).

formulae for the normal form coefficients are derived explicitly by [Planeaux and Jensen, 1986] and [Nayfeh, 1993]. Next we outline the procedure of the normal form method briefly. To apply this method we need to transform the steady states to the origin by the translation

$$u_1 = x - x_{ss}, \quad u_2 = y - y_{ss}, \quad u_3 = z - z_{ss}$$

210

Under this transformation the model becomes

$$
\begin{pmatrix} \dot{u}_1 \\ \dot{u}_2 \\ \dot{u}_3 \end{pmatrix} = \begin{pmatrix} -\dfrac{1}{\theta} - y_{ss}^2 & -2x_{ss}y_{ss} & 0 \\[2mm] y_{ss}^2 & -\dfrac{1}{\lambda} - \dfrac{1}{\theta} + 2x_{ss}y_{ss} - \kappa & \dfrac{1}{\lambda} \\[2mm] 0 & \dfrac{1}{\rho\lambda} & -\dfrac{1}{\rho\lambda} - \dfrac{\sigma}{\rho\theta} - \kappa \end{pmatrix}
$$

$$
\begin{pmatrix} u_1 \\ u_2 \\ u_3 \end{pmatrix} + \begin{pmatrix} -2y_{ss}u_1u_2 - x_{ss}u_2^2 - u_1u_2^3 \\[2mm] 2y_{ss}u_1u_2 + x_{ss}u_2^2 + u_1u_2^3 \\[2mm] 0 \end{pmatrix} \tag{8.15}
$$

Equivalently, the model can be written as

$$
\dot{u} = Au + F(u) \tag{8.16}
$$

This model always has a trivial solution $u = 0$. We will express the results of this latter model in term of x_{ss} as its bifurcation parameter. The original steady state equations are used to express y_{ss}, z_{ss} and θ in term of x_{ss} and the remaining parameters. Figure 8.20 shows that the saddle-node bifurcation for the original model becomes transcritical bifurcation for the new model.

The second step is to transform the Jacobian matrix close to the F_2 point into Jordan canonical form. Using the transformation

$$
u = Cv
$$

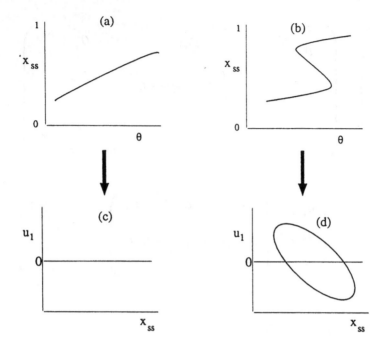

Figure 8.20. Bifurcation diagrams: (a), (b) original model (c), (d) modified model.

the linearized matrix becomes

$$J = C^{-1} AC = \begin{pmatrix} \mu + i\omega & 0 & 0 \\ 0 & \mu - i\omega & 0 \\ 0 & 0 & v \end{pmatrix}$$

and the model equation becomes

$$\dot{v} = Jv + G(v) \tag{8.17}$$

where $G(v) = C^{-1} F(Cv)$. The matrices C and C^{-1} consist of the right and the left eigenvectors of the eigenvalues $(\mu \pm i\omega, v)$ respectively.

The next step is to use an additional coordinate change to simplify the nonlinear part, $G(v)$. We introduce

$$v = w + P(w) \tag{8.18}$$

to arrive to the normal form

$$\dot{w} = Jw + \Phi(w) \tag{8.19}$$

where $P(w)$ is a polynomial expansion, beginning with quadratic terms of the same form as G. The derivative of equation 8.18 has the form

$$\dot{v} = (I + DP)\dot{w} = Jv + G(v) \tag{8.20}$$

or

$$Jw + \Phi + DPJw + DP\Phi = Jw + JP + G \tag{8.21}$$

which yields

$$\Phi = JP - DPJw - DP\Phi + G \tag{8.22}$$

The nonlinear quantities can be written as

213

$$G = G_2 + G_3 + \ldots$$

$$P = P_2 + P_3 + \ldots$$

$$\Phi = \Phi_2 + \Phi_3 + \ldots$$

where the terms G_2, P_2, and Φ_2 refer to the quadratic terms, and G_3, P_3, and Φ_3 are the cubic terms. Equation 8.22 relate the coefficients of the monomials P, Φ, and G. The coefficients of P are chosen so that the right hand side of equation 8.22 vanishes whenever possible. Equation 8.22 is solved by equating the coefficient of like powers. For quadratic terms, we obtain

$$\Phi_2 + DP_2Jw - JP_2 = G_2$$

and for cubic terms

$$\Phi_3 + DP_3Jw - JP_3 = G_3 - DG_2P_2 - DP_2\Phi_2$$

So it is clear that the linear operator, known as *Lie bracket*,

$$JP - DPJw$$

plays a very important role in this analysis. Monomials in the null space of this operator give rise to nonvanishing nonlinear terms in Φ. The details of the steps followed above are described in the Appendix of Planeaux and Jensen [1986] and in [Nayfeh, 1993; Guckenheimer and Holmes, 1983; Wiggins, 1990]. The resulting $\mathbf{F_2}$ normal form is expressed in cylindrical coordinates ($v_1 = re^{i\Theta}$, $v_2 = re^{-i\Theta}$, $v_3 = z$):

214

$$\dot{r} = \mu r + a_1 rz + a_2 r^3 + a_3 rz^2$$

$$\dot{z} = va + b_1 r^2 + b_2 z^2 + b_3 r^2 z + b_4 z^3$$

$$\dot{\Theta} = \omega + \dots \tag{8.23}$$

Note that Θ does not appear in the normal form of the model. As a result of that, the problem has been reduced to the study of a two-dimensional model. In the normal form model, the eigenvalues and the coefficients depend on the model parameters. The definitions of these coefficients are taken from [Planeaux and Jensen, 1986]. The nonlinear terms of our model can be expressed as

$$G(v) = \begin{pmatrix} g(v) \\ \tilde{g}(v) \\ h(v) \end{pmatrix}$$

where

$$g(v) = \sum_{m+n+r=2}^{3} \frac{g_{mnr}}{m!n!r!} v_1^m v_2^n v_3^r$$

$$h(v) = \sum_{m+n+r=2}^{3} \frac{h_{mnr}}{m!n!r!} v_1^m v_2^n v_3^r$$

215

The coefficients are defined to be:

$$a_1 = Re(g_{101})$$

$$b_1 = h_{110}$$

$$b_2 = 0.5h_{002}$$

$$a_2 = 0.5Re(g_{210}) - \frac{1}{2\omega} (Im\,(g_{200}g_{110}) - 0.5Im\,(g_{011}h_{200}))$$

$$a_3 = 0.5Re(g_{102}) - \frac{1}{2\omega} (Im\,(g_{200}g_{002}) - Im\,(g_{110}\tilde{g}_{002}) - 2Im\,(g_{002}h_{101}))$$

$$b_3 = h_{111} - \frac{1}{\theta} (Im\,(h_{200}g_{011}) + 2Im\,(h_{101}g_{110}))$$

$$b_4 = \frac{h_{003}}{6} - \frac{1}{\omega} (Imh_{101}g_{002}) \tag{8.24}$$

There are three steady state solutions of the normal form. The first two of these solutions correspond to steady states of the original model. There are given as

$$r = 0, \quad z = 0$$
$$r = 0, \quad z = \frac{-v}{b_2} + O(2)$$

while the third solution ($r \neq 0, z$) corresponds to the periodic branch. Torus bifurcations are Hopf bifurcations of the planar normal form. We illustrate the solution structure for the normal form model by studying the model around the $\mathbf{F_2}$ point:

$$\sigma = 0.1, \quad \lambda = 20.0, \quad \rho = 20.0, \quad \kappa = 0.00445, \quad \beta = 1.0$$

For this case, the bifurcation diagram in term of the original model is given in Figure 8.17. The three steady solutions of the normal form model are shown in Figure 8.21. Note that the Hopf point, where the solution ($r \neq 0, z \neq 0$) emerges at, lies closely to the turning point. Also it can be seen that along the periodic solutions, the model exhibits a secondary bifurcation leading to tori. Figure 8.22 shows the trajectories for the case $\theta = 156.5417$, $x_{ss} = 0.5420012$ using the modified model and the normal form model. The simulations of both models show that the models exhibit similar quasiperiodic motions. For the same example, Figure 8.23 shows the saddle-node, Hopf, torus singularities in the κ, θ-plane. It shows that the torus bifurcation first appears exactly at the $\mathbf{F_2}$ point. Using the normal form model, the stability and the direction of the torus branches can be examined similarly to studying the properties of the periodic solutions emanating from a Hopf point. We believe that there is still a lot of work that can be done in analyzing the $\mathbf{F_2}$ degeneracies to a greater depth than we have reached in this section.

8.4 Codimension-two Singularities

8.4.1 $\mathbf{H_{02}}$: Multiple Hopf Points

The $\mathbf{H_{02}}$ singularity defines the conditions for Hopf point multiplicity. The conditions for the $\mathbf{H_{02}}$ singularities are:

$$F = F_x = F_{xx} = 0, \quad F_{xxx} \neq 0$$

where F is defined by equation 8.7. H_{02} degeneracies have been identified for cases 3, 4, 6, 7, 8 and 9 in Section 8.2. The existence of H_{02} degeneracies in these cases have led to more new, interesting bifurcation patterns.

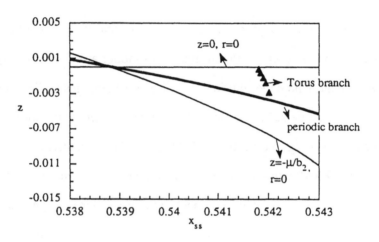

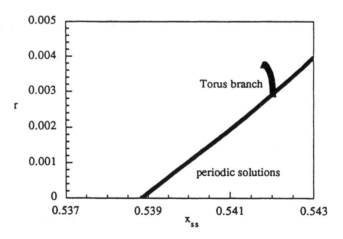

Figure 8.21. Steady state structure of the normal form model.

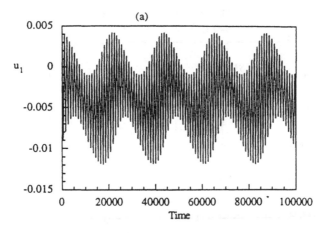

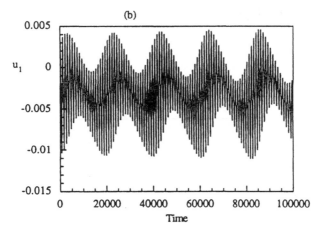

Figure 8.22. System trajectory for the case $\theta = 156.5417$, $\kappa = 0.00445$, $\beta = 1.0$, $\sigma = 0.1$, $\lambda = 20.0$, $\rho = 20.0$ (a) normal form model, (b) modified model.

We examine the Hopf points curves for two cases. First, we consider the case shown in Figure 8.5, the top right hand corner of which is enlarged in Figure 8.24 to show a part of the branch set where only unique behaviour can be found. Three Hopf points curve are shown in the θ, κ-plane. Unique Hopf points are obtained for β values

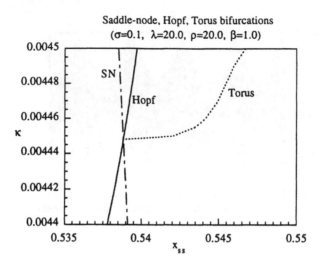

Figure 8.23. Saddle-node, Hopf, torus singularities for the case β = 1.0, σ = 0.1, λ = 20.0, ρ = 20.0.

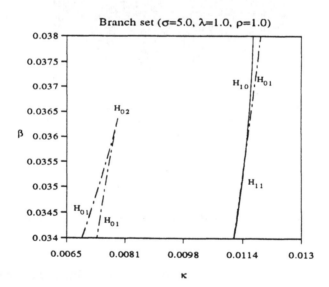

Figure 8.24. Branch set: σ = 5.0, λ = 1.0, ρ = 1.0.

220

-ying above its value at **H₀₂** point while S-shape Hopf point curves are found for β values -ying below the **H₀₂** point, (Figure 8.25). In the second example, we consider the Hopf points around the **H₀₂** singularity found for the case ($\sigma = 8.0,\ \lambda = 9.0,\ \rho = 1.0$), Figure 8.13). An isola of Hopf points is identified for this case, see Figure 8.26.

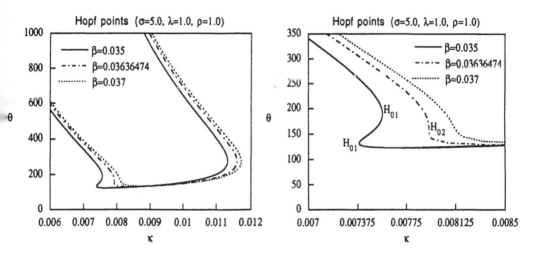

Figure 8.25. Hopf points multiplicity: $\sigma = 5.0,\ \lambda = 1.0,\ \rho = 1.0$.

The **H₀₂** conditions can be used to define the σ, λ, ρ values at which the model has branch sets with **H₀₂** degeneracies. We use the **H₀₂** conditions together with the steady state relationships to solve for x_{ss} and y_{ss} and the parameters θ, κ, and β in terms of σ. The **H₀₂** curves are calculated for two values for ρ and three values for λ. Figure 8.27 shows the **H₀₂** curve for the case $\rho = 1.0$. It can be seen that for low flow-rate ratio, σ, the model does not show any **H₀₂** points. Similar observation can be made when looking at the case $\rho = 4.0$ shown in Figure 8.28. Note also that when λ increases, the model exhibits **H₀₂** degeneracies for lower σ values.

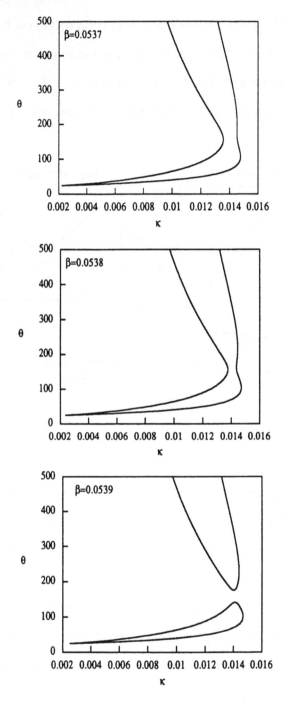

Figure 8.26. Hopf points multiplicity: $\sigma = 8.0$, $\lambda = 9.0$, $\rho = 1.0$.

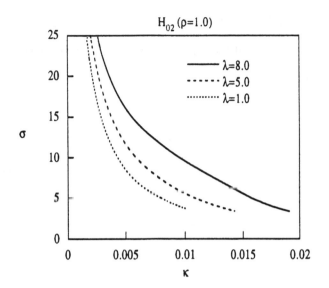

Figure 8.27. **H₀₂** curves $\rho = 1.0$.

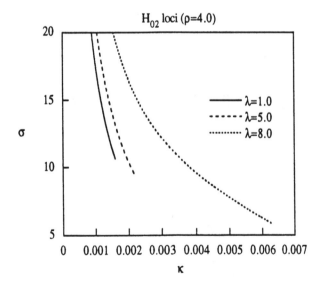

Figure 8.28. **H₀₂** curves $\rho = 4.0$.

8.4.2 H₂₀: Multiple Periodic Solutions

These types of degeneracies lead to two limit points on the periodic branch and three periodic solutions. The formula of Golubitsky and Langford are used for identifying these singularities in the parameter space. For some of the cases in Section 8.2 the H_{20} points are found on the lower H_{10} curves which lie inside the multiplicity region. Drawing the bifurcation diagrams around the H_{20} is not easy. The reason is that the homoclinic bifurcation occurs before the periodic branch shows a complete hysteresis shape.

We show the bifurcation diagrams in the vicinity of the H_{20} for the case $\sigma = 1.0$, $\lambda = 1.0$, $\rho = 2.0$. Figure 8.29 shows the branch set for this case. In Figure 8.30, we show the bifurcation patterns for β values at the left of the H_{20} point where the model exhibits multiple periodic solutions (Figure 8.30a,b,c,d). While Figure 8.30e, f, and g show the bifurcation diagrams for β values to the right of the H_{20} point where unique periodic solutions are obtained.

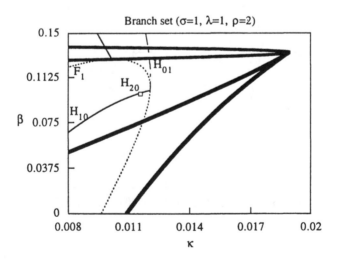

Figure 8.29. Branch set for $\sigma = 1.0$, $\lambda = 1.0$, $\rho = 2.0$.

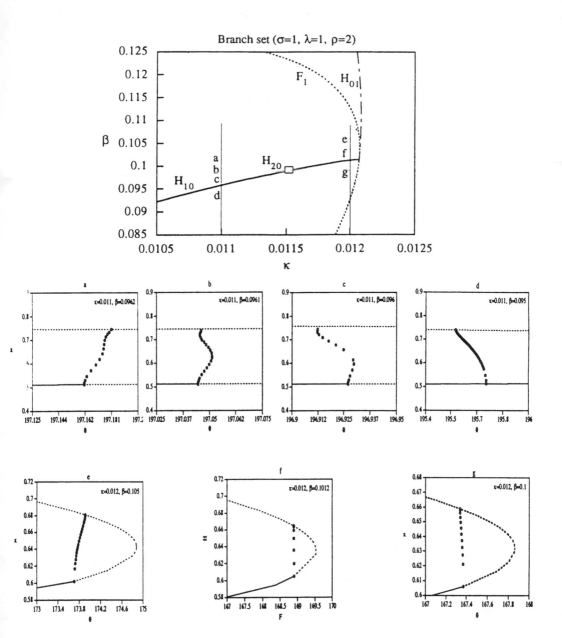

Figure 8.30. Bifurcation diagrams around $\mathbf{H_{20}}$.

225

8.4.3 H_{11}: Isolated Periodic Solutions

H_{11} occurs when the H_{01} curve intersects H_{10} tangentially. One form of H_{11} degeneracies has been observed in six cases in Section 8.2. H_{01} are found in the unique regions of the cases 3, 4, 6, 7, 8, 9. These tangential intersections agree with the unfoldings constructed by Golubitsky and Langford using singularity theory. A third curve emerging from the H_{11} corresponds to the birth of an isola of limit cycles. On the branch set of the case ($\sigma = 5.0, \lambda = 1.0, \rho = 1.0$), we define the region of the isolated periodic solutions in Figure 8.31. In the same figure, we show isolated limit cycles by examining the bifurcation diagrams for $\beta = 0.029$ for three values for κ.

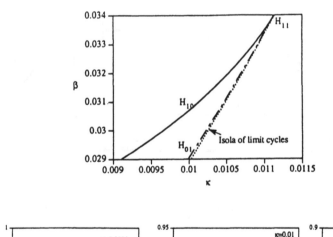

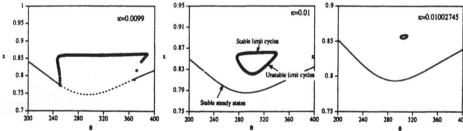

Figure 8.31. H_{11} degeneracy for the case $\sigma = 5.0, \lambda = 1.0, \rho = 1.0$.

226

Chapter 9

CHAOS - THEORY

Up to this point the oscillations we have considered have been periodic, but one frequently encounters irregular oscillations like those illustrated in Figure 9.1. This irregularity is part of the intrinsic dynamics of the system and not the result of unpredictable outside influences as in stochastic systems. Chaos' central characteristic is that it is enormously and ineluctably sensitive to initial conditions. In Figure 9.1, we can observe that the amplitude is bounded and the frequency seems to vary only slightly. In the first section, we discuss the structure of chaotic attractors. Some routes to such behaviour are presented in the second section. A full discussion of the characteristics of

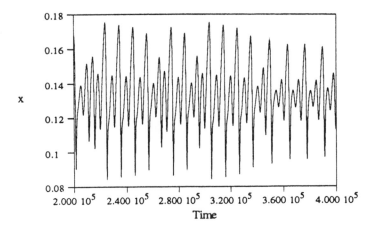

Figure 9.1. Chaotic motions.

and routes to chaos are given in several books (see, for example, the references [Baker and Gollub, 1990; Buckenheimer and Holmes, 1983; Hilborn, 1994; Marek and

Schreiber, 1991; Moon, 1992; Parker and Chau, 1986; Seyedel, 1988; Scott, 1991; Wiggins, 1990, 1988]).

9.1   Deterministic Chaos

The most common definition of chaos arises from a concept based on the behaviour of trajectories in phase space. This concept is based on an important property of dissipative systems: the contraction of volume. Consider the following dynamic system

$$\frac{dx}{dt} = F(x, \theta) \tag{9.1}$$

Where x is the state variable vector and θ is the bifurcation parameter. First assume that initially all trajectories are bounded by a subset A in IR^n and let V be its volume. This set is generally deformed under the time evolutions according to equation 9.1. The time evolution of the volume V of the set A is given according to *Liouville's theorem* [Marek and Schreiber, 1991; Baker and Gollub, 1990] as

$$\frac{dV}{dt} = \int_A divF(x(t))dx \tag{9.2}$$

The global contraction of subsets of the state space will be guaranteed if

$$divF(x(t)) < 0$$

for all x. Asymptotic solutions occur on sets which have zero volume. An attractor is defined to be a proper subset of phase space IR^n to which all trajectories converge as time ncreases without bound. The structure of an attractor can be simple, for example, a point n the state space corresponds to stationary behaviour and a closed curve to periodic olutions. However, an overall contraction of volumes does not exclude complicated dynamics. The trajectory set may be expanding in some directions in the state space even f its volume vanishes asymptotically. This may cause the folding of different parts of A apon itself under time evolution and the asymptotic set may then have very complicated structure as well as complicated dynamics. The Smale horse-shoe, a one-to-one discrete system that maps the unit square into itself, has been used to give sufficient conditions for the existence of chaotic set in dynamic systems . We will not discuss the dynamics of the Smale horse-shoe here but refer the reader to the references [Guckenheimer and Holmes, 1983; Wiggins, 1990].

In the case of a chaotic "strange" attractor, trajectories tend to diverge, two different initial conditions, no matter how close together, do not have similar histories. They begin to move apart, the initial separation grows exponentially, and eventually become completely uncorrelated. This implies very important property for these attractors: chaos in deterministic systems implies a sensitive dependence on initial conditions.

9.1.1 Lyapunov Exponents

The average rate of divergence of neighboring trajectories on the attractor is quantified through the use of Lyapunov exponents. The direction of maximum divergence or convergence is a changing local property of the strange attractor. The motion must monitored at each point along the trajectory. Therefore, a small sphere is defined whose center is a given point in the attractor and whose surface consists of points from the nearby trajectories. As the center of the sphere and its surface points evolve in time, the sphere becomes an ellipsoid, with principle axis in the direction of contraction and expansion, Figure 9.2 [Baker and Gollub, 1990]. The average rate of expansion or contraction along the principle axes are the Lyapunov exponents. The definition for Lyapunov exponents is defined as follows [Parker and Chau, 1986]. First, we need to

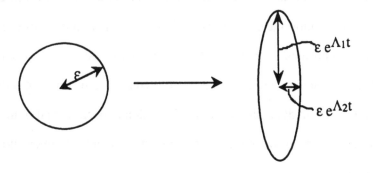

Figure 9.2. Principal axes in the directions of expansion and contraction at a point in the trajectory.

define the variational equation which is a matrix-valued time-varying linear differential equation. Consider the nth order system

$$\dot{x} = F(x) \tag{9.3}$$

with solutions ϕ, that is

$$\dot{\phi}_t(x_0, t_0) = F(\phi_t(x_0, t_0)), \quad \phi_{t_0}(x_0, t_0) = x_0 \tag{9.4}$$

Differentiation equation 9.4 with respect to x_0 to obtain

230

$$D_{x_o} \dot{\phi}_t (x_o, t_o) = D_x F (\phi_t (x_o, t_o)) D_{x_o} \phi_t (x_o, t_o), \quad D_{xo} \phi_{to} (x_o, t_o) = I \qquad (9.5)$$

Define $\Phi_t (x_o t_o) = D_{x_o} \phi_t (x_o, t_o)$. Then, equation 9.5 becomes

$$\dot{\Phi}_t (x_o, t_o) = D_x F (\phi_t (x_o, t_o)) \Phi_t (x_o, t_o), \quad \Phi_{to} (x_o, t_o) = I \qquad (9.6)$$

which is the variational equation. For any initial condition $x_o \in IR^n$, let $m_1 (t), ..., m_n (t)$ be the eigenvalues of $F_t (x_0)$. The Lyapunov exponents of x_o are defined to be [Parker and Chau, 1986]:

$$\Lambda_i = \lim_{t \to \infty} \frac{1}{t} \ln |m_i (t)|, \quad i = 1, ..., n \qquad (9.7)$$

whenever the limit exists. They are average rates of expansion (or contraction) if $\Lambda_i > 0 (\Lambda_i < 0)$.

Lyapunov exponents are a generalization of the eigenvalues at fixed points and of Floquet multipliers for limit cycles. A nonchaotic hyperbolic attractor is asymptotically stable if all its Lyapunov exponents are negative. For any attractor in an autonomous system except an equilibrium point, one Lyapunov exponent is always zero. In dissipating systems, it is required for stable attractor that the sum of the Lyapunov exponents be negative. At least one Lyapunov exponent of a chaotic attractor is positive; it accounts for the sensitive dependence on initial conditions. From these facts, it follows that a strange attractor must have at least three Lyapunov exponents. Thus, chaos can not occur in first or second order autonomous systems and at least three variables are required for chaos to exist. The three space is sufficient to to allow for the divergence of trajectories, confinement of the motion to a finite region of the phase space of the dynamical system and for the uniqueness of trajectory. In three dimensional systems, the only possibility is $\Lambda_1 > 0$, $\Lambda_2 = 0$ and $\Lambda_3 < 0$. Since contraction must outweigh expansion, a further condition on stable three-dimensional chaos is $\Lambda_3 < -\Lambda_1$. The Lyapunov exponents of the different types of attractors are summarized in Table 9.1 [Parker and Chau, 1986]. Several algorithms for estimating all n Lyapunov exponents of any asymptotically stable attractors are to be found in the literature: see for example, Parker and Chau [1986], McKarnin, Schmidt, and Aris [1988];Wolf, [1984].

Attractor Type	Poincaré Map	Lyapunov Exponents	Dimension
Steady State		$0 > \Lambda_1 \geq ... \geq \Lambda_n$	0
Periodic	one or more points	$\Lambda_1 = 0, \, 0 \geq \Lambda_2 \geq ... \geq \Lambda_n$	1
Two-periodic Torus	one or more closed curves	$\Lambda_1 = \Lambda_2 = 0, \, 0 > \Lambda_3 \geq ... \geq \Lambda_n$	2
Chaotic	fractal-like	$\Lambda_1 > 0, \, \sum \Lambda_i < 0$	noninteger

Table 9.1. Lyapunov exponents.

9.1.2 Poincaré Map and Return Map

Another useful description of the dynamics is to look only at the point of penetration of the trajectory on a fixed transverse plane Σ. This is known as a Poincaré map [Hilborn, 1994; Moon, 1992; Parker and Chau, 1986]. Consider the motion as a trajectory in a three-dimensional phase space (x, y, z). A Poincaré map can be defined by constructing a two dimensional oriented surface in this space and looking at the points (x_n, y_n, z_n) where the trajectory crosses this surface. For example, one can choose a plane $n_1x + n_2y + n_3z = c$ with normal vector $\mathbf{n} = (n_1, n_2, n_3)$ tangential to the solution curve. The appearance of fixed points and closed orbits on the Poincaré map indicates periodic solutions. A single fixed point of the map corresponds to a period-one solution, two fixed points corresponds to period-2 solution and k-fixed points corresponds to period-k solutions. Sometimes the map becomes a continuous closed curve. The motion in this case is called quasiperiodic or motion on a torus. If the Poincaré map does not consist of either a finite set of points or a closed curve, the motion may be chaotic. Here the map may consist of open curve or fuzzy collection of points. If the Poincaré map consists of an open curve which can be considered as a set of points along some one-dimensional line. This suggests that, if one could parameterize those points along the line by a variable x, it would be possible to define a function which relates x_{n+1} and x_n,

$$x_{n+1} = f(x_n)$$

This function is called return map. Here, one samples some dynamic variable using a Poincaré section as discussed above, then plots each x_n against its successor value x_{n+1}. This technique is used to model the dynamics of some complex physical system as a one-dimensional map.

For many cases, one can enlarge a portion of the Poincaré map or return map and observe further structure. If this structured set of points continues to exist after several enlargements then the map is said to be *fractal-like*, and one says the motion behaves as a chaotic attractor. So the appearance of fractal-likeness in these maps is a strong indicator of chaos [Moon, 1992].

9.1.3 Power Spectra

The power spectra allow us to differentiate between periodic and chaotic behaviour. It can be computed by the standard software available on most computers based on the Fast Fourier Transform method. The result of applying the power spectrum on any time series $(y_k(0), y_k(\Delta t), y_k(2\Delta t) \dots)$ consists of a sequence of frequencies with associated amplitudes. This set of numbers is presented graphically, displaying amplitude versus frequency. This method is illustrated qualitatively in the following four cases. First, we show the power spectrum of a periodic time series (Figure 9.3a). The fundamental frequency f, and its harmonics $2f$, $3f$, ... are visible as sharp peaks. Figure 9.3b shows power spectrum of the period doubling case. Subharmonics occur with the frequency $f/2$ and integer multiples. Similarly a period-4 shows peaks with distance $f/4$. The power spectrum resembling in Figure 9.4 is for the case of the quasiperiodic motions resulted from the bifurcation of periodic solutions on a torus. It shows another fundamental

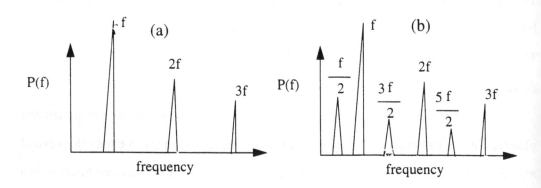

Figure 9.3. Power spectrum for periodic motions.

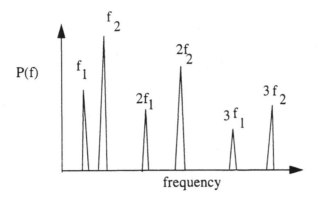

Figure 9.4. Power spectrum for quasiperiodic motions.

frequency f_2 which is not commensurable with f_1. Both periodic and quasiperiodic motions are characterized by power spectra with small amplitude noise and discrete peaks. This in contrast to the chaotic case, which is characterized by a broad-band noise (Figure 9.5).

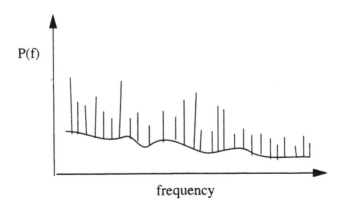

Figure 9.5. Power spectrum for chaotic motions.

9.2   Routes to Chaos

The theory of nonlinear dynamical systems is not able to say in advance when and how a dynamic system will exhibit chaotic behaviour. There are several routes to chaos; these transitions can be gathered into two large categories [Hilborn, 1994]. In the first, local routes, periodic motions occur for a range of parameter values. As the bifurcation parameter is varied, the periodic behaviour loses its stability and chaotic behaviour appears. In the second category, the transition is not marked by any change in the equilibrium points or the limit cycles of the system but it is due the interaction of trajectories with various attractors in the state space [Guckenheimer and Holmes, 1983; Wiggin, 1988].

9.2.1   Local Bifurcations

Period-Doubling Route to Chaos

This route has been observed in many chemical systems [Cordonier, Schmidt and Aris, 1990; Lynch, 1992a,b; McKarnin, Schmidt and Aris, 1988; Peng, Scott and Showalter, 1990] and indeed is considered to be one of the principle routes to chaotic behaviour in any context [Hilborn, 1994; Moon, 1992]. Period-doubling bifurcation occurs when one of the Floquet multipliers crosses the unit circle at -1 on the real axis, while the other multipliers remain inside the unit circle.

In this phenomena, one starts with a system with a periodic motion. Then as some bifurcation parameter, θ, is varied the system undergoes a secondary bifurcation to a periodic motion with twice the period of the original solution. By changing the parameter θ further, the system bifurcates to periodic solution with four times the period of the first periodic oscillation. Finally, successive period-doubling take place leading to chaotic behaviour. One feature of this route is that the critical value of θ at which successive period doubling bifurcations occur obey the following rule

$$\frac{\theta_n - \theta_{n-1}}{\theta_{n+1} - \theta_n} \rightarrow \delta = 4.6692016$$

as $n \rightarrow \infty$. This number is called the Feigenbaum number. The bifurcation values accumulate at a critical value for the which the system becomes chaotic. The band of parameter values for which the chaotic motion exists may be of finite width.

Quasi-periodic Route to Chaos

This route to chaos based on torus bifurcation was proposed by Newhouse, Ruella and Takens [1978]. In this case, only two bifurcations take place before the onset of chaos. First, the system undergoes Hopf bifurcation. The second transition is the 2-torus bifurcation, which means that two coupled limit cycles are present. When the frequencies of these oscillations are not commensurate, the observed motion itself is not periodic but is said to be quasiperiodic. As shown in Table 9.1, the Poincaré map of quasiperiodic motion is a closed curve in the phase plane. If the the frequencies are commensurate, the trajectory on the torus is close. In this case the Poincaré map becomes a set of points generally arranged in a circle. Chaotic motions are observed in such cases when the system undergoes another bifurcation so that three simultaneous coupled limit cycles are present. But this kind of bifurcation is not generic and chaos is more likely than bifurcation to a 3-torus.

Intermittency

In a third route to chaos, one observes long periods of regular motion with interrupted by bursts of aperiodic oscillations of finite durations. As some bifurcation parameter of the system is varied, the chaotic bursts become longer and occur more frequently until the entire time series becomes chaotic. This route has been observed in several chemical models [Jorgenson and Airs, 1983; Lynch, 1992b, 1993]. The theory of intermittency has been based on the study of iterated mapping. An explanation of such behaviour has been posited in terms of one-dimensional maps or difference equations [Lynch, 1993]. A full discussion is given in the book by Hilborn [1994], where three types are recognized and distinguished based on the change of the Floquet multipliers at the bifurcation point.

9.2.2 Global Bifurcations

The theory of this scenario is less well developed than the theory for period-doubling, quasi-periodicty, and intermittency. The transition is due to the interaction of trajectories with various unstable steady-state points and limit cycles in the state space. The common features are the so-called homoclinic orbits and hetroclinic orbits. These orbits may suddenly appear as the bifurcation parameter is changed. References that give a detailed mathematical treatment of the global chaotic behaviour are Guckenheimer, Holmes, [1983]; Hilborn [1994]; and Wiggins [1990, 1988] . An example of homoclinic situation is given in the following section.

Homoclinicity

A general example of a route to chaos is due to Shil'nikov [1970]. It applies to a steady-state point of saddle-focus type, when there is a homoclinic trajectory passing through the

238

saddle-focus. The theorem says roughly that if the only real eigenvalue and the real part of a pair of complex eigenvalues of the linearized system are of opposite sign and the absolute value of the former is larger, then there exists a set of chaotic trajectories near the original homoclinic trajectory.

Figure 9.6 shows a representation of a Shil'nkov homoclinicity. The unstable focal plane forms a slow manifold on which the system unwinds. As the system trajectories lift off this plane, it is ultimately reinjected along the strongly attracting stable manifold to the steady-state point. When the structure is perturbed, a strange attractor may form with the initial point for each slow unwinding depending sensitively on the

rapid reinjection process. The simplest form for a model that can satisfy this scenario is the three-variable system:

$$\frac{dx}{d\tau} = vx - \omega y + f_1(x, y, z)$$

$$\frac{dy}{d\tau} = \omega x + vy + f_2(x, y, z)$$

$$\frac{dz}{d\tau} = \mu z + f_3(x, y, z) \tag{9.8}$$

The Shil'nkov homoclinicity conditions are equivalent to $-\mu \gg v > 0$.

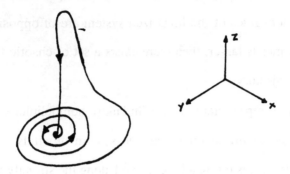

Figure 9.6. Shil'nikov homoclinicity.

Chapter 10

CHAOS - MODEL I

The mathematical model for the first case, coupling through the reactant, is

$$\frac{dx}{d\tau} = \frac{z - x}{\lambda} - \frac{x}{\theta} - xy^2$$

$$\frac{dy}{d\tau} = \frac{\beta - y}{\theta} + xy^2 - \kappa y$$

$$\rho\frac{dz}{d\tau} = -\frac{z - x}{\lambda} + \frac{\sigma}{\theta}(1 - z) \qquad (10.1)$$

This simple model fulfills the necessary condition of having a minimum of three ordinary differential equations for the existence of chaotic behaviour. A numerical examination of the first model reveals that chaotic oscillations are possible. It is found that period-doubling occurs as the model behaviour changes from a stable periodic to a chaotic motion. This behaviour is studied using bifurcation diagrams, phase plane, Poincaré map, Lyapunov exponents, and power spectrum.

10.1 Period-Doubling in the Parameter Space

It is shown that period-doubling bifurcation is generic in one-parameter families of one-dimensional maps, and it persists if the underlying model undergoes small changes. Therefore, the codimension of period-doubling bifurcation is zero. The Floquet multipliers are used for the detection of period-doubling bifurcation. Therefore, leaving or entering the unit circle by a Floquet multiplier equal to -1 signals a potential period-doubling bifurcation. A continuation subroutines such as Auto [Doedel and Kerevez, 1981] or Cont [Marek and Schreiber, 1991] can be used to monitor the Floquet multipliers along the periodic solutions branches to locate the critical bifurcation parameter at which one of the multipliers crosses the unit circle at -1. The results are checked by using Gear's method for integration of stiff systems of ordinary differential equations. This techniques is used to define the locus of period-doubling bifurcation in the parameter space. We start with the parameter values

$$\kappa = \frac{4}{450}, \quad \beta = \frac{4}{15}, \quad \lambda = 225, \quad \sigma = 2, \quad \rho = 20$$

For these parameter values, the model exhibit `unique' steady state patterns. We fix four of these parameters and vary the fifth parameter to define the period-doubling bifurcation in the parameter space defined by this chosen parameter and the bifurcation parameter θ. Figure 10.1 shows the Hopf points curve and the period doubling points for three cases. In the first case we vary λ and solve for the bifurcation parameter value at which period-doubling occurs. In the second and the third cases, we vary the parameters σ and ρ, respectively. The period-doubling point is located close to the first Hopf point (at the smallest residence time). In all these cases the period-doubling loci have a minimum value. Below these values the system does not undergo a secondary bifurcation. For all the values at which the system has period-doubling bifurcation, the system has a unique steady state. In general, period-doubling does not imply chaos, we illustrate this by studying the case: $\kappa = \frac{4}{450}, \beta = \frac{4}{15}, \lambda = 225, \sigma = 2, \rho = 5$. For these parameter values, Figure 10.1c shows that the system undergoes a period-doubling bifurcation. The bifurcation diagram for this case is given in Figure 10.2. We can see

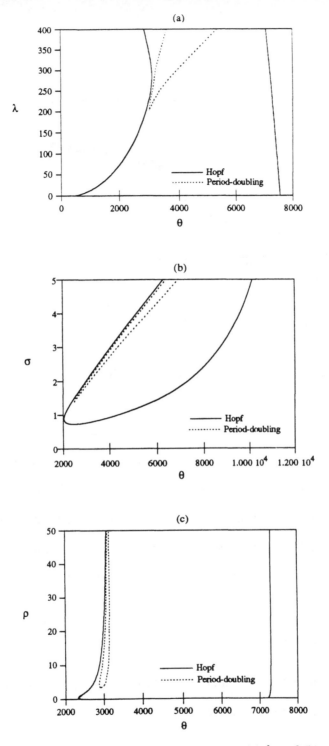

Figure 10.1. Period-doubling loci in the parameter space (a) λ vs θ (b) σ vs θ (c) ρ vs θ.

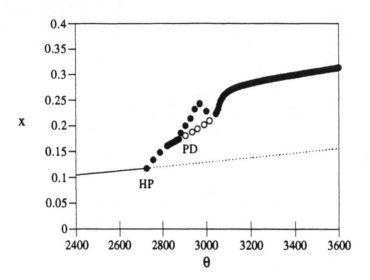

Figure 10.2. Bifurcation diagram for the case $\kappa = 4/450$, $\beta = 4/15$, $\lambda = 225$, $\sigma = 2$, $\rho = 5$.

that with the increase of the value of the bifurcation parameter θ above 2700 we observe one periodic orbit. Next, period-doubling occurs at $\theta \approx 2880.6$ after which the system has period-2 oscillations. Increasing the bifurcation parameter further does not lead to period-4 but at $\theta \approx 3040.0$ it returns the system back to period-one oscillations. The time series for five cases taken from the period-doubling band are shown in Figure 10.3.

10.2 Chaotic Behaviour

We have, however, found that for higher values of ρ the system can display chaotic behaviour through period-doubling route. In particular, for the case

244

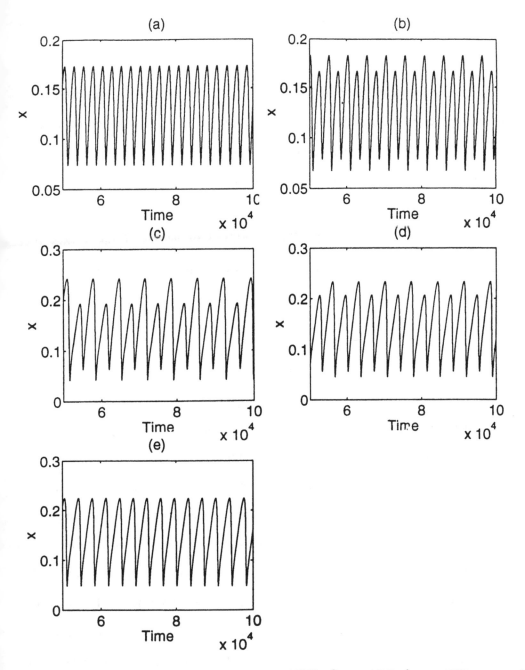

Figure 10.3. Time series for the case $\kappa = 4/450$, $\beta = 4/15$, $\lambda = 225$, $\sigma = 2$, $\rho = 5$, (a) $\theta = 2870$, (b) $\theta = 2880$, (c) $\theta = 3010$, (d) $\theta = 3030$, (e) $\theta = 3045$.

$$\kappa = \frac{4}{450}, \quad \beta = \frac{4}{15}, \quad \lambda = 225, \quad \sigma = 2, \quad \rho = 20$$

we observe infinite period-doubling bifurcations leading to chaos. The bifurcation diagram in Figure 10.4 shows that as the bifurcation parameter is changed beyond the value $\theta = 3035.5243$, the model undergoes a period-doubling bifurcation. As θ is changed further the model bifurcates to period-4, then period-8 and so on. We expect that if the parameter θ is changed beyond some critical value, then chaotic motion exists. First we look at a scaling property of period-doubling bifurcations. If one knows three consecutive values of the period doubling bifurcation parameter θ, θ_{n-1}, θ_n, θ_{n+1}, then the Feigenbaum number can be found:

$$\delta = \lim_{t \to \infty} \frac{\theta_n - \theta_{n-1}}{\theta_{n+1} - \theta_n} = 4.6692016$$

Feigenbaum proposed that in physical systems that undergoes such period-doubling phenomena, the above limit δ should hold. Our computations have shown the validity of this universal number for our model. We summarize these computations in Table 10.1. The first column in Table 10.1 shows the calculated values for six consecutive period-doubling bifurcation parameter. These values are used to estimate the scaling number δ in the second column. It is found approximately equal to the proposed value. The bifurcation values θ_n converge in an exponential way as we approaches the limiting value θ_∞ at which chaotic dynamics develop. This limiting value is estimated in the third column of Table 10.1 using the number $\delta = 4.6692016$. We will study the behaviour of the model in the period-doubling regions using different tools in the following sections.

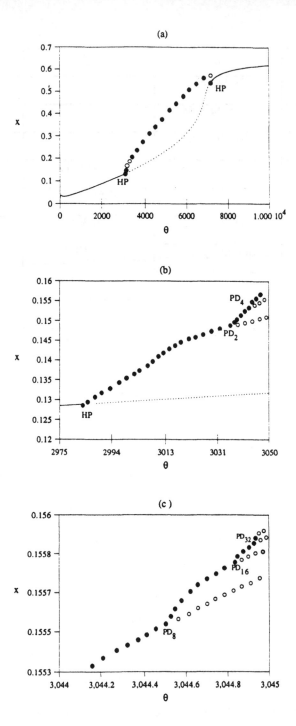

Figure 10.4. Period-doubling bifurcation leading to chaos (κ = 4/450, β = 4/15, λ = 225, σ = 2, ρ = 20).

Bifurcation Type	θ	δ	θ^*
Hopf	2983.245		
Period-2	3035.241654		
Period-4	3042.919485		
Period-8	3044.508503	4.8318112	
Period-16	3044.846948	4.695055	3044.8488221
Period-32	3044.919218	4.683064	3044.9217079
Period-64	3044.934395	4.761810	3044.9373178
Period-128		4.6692016	3044.9406610
...	...	...	...
θ_∞		4.6692016	3044.9415721

Table 10.1. Scaling property of period-doubling bifurcation.

10.2.1 Lyapunov Exponents

Chaos in dynamics implies a sensitivity of the outcome of the dynamical system to changes in initial conditions. Sensitive dependence occurs in a flow with an expanding component. Since Lyapunov exponent indicates expansion, what distinguishes chaotic attractors from regular attractors is the existence of a positive Lyapunov exponents. In a three-dimensional systems, the only possibility is (+, 0, -), that is $\Lambda_1 > 0$, $\Lambda = 0$ and $\Lambda_3 > 0$. A further condition on stable three-dimensional chaos is $\Lambda_3 < \Lambda_1$.

Different approaches for estimating Lyapunov exponents are given by [Marek and Schreiber, 1991; Parker and Chau, 1986; Shimada and Nagashima, 1979; Wolf, 1984]. The calculations of the largest Lyapunov exponents of the first model as a function of the residence time θ is given in Figure 10.5. It shows that the chaotic motion

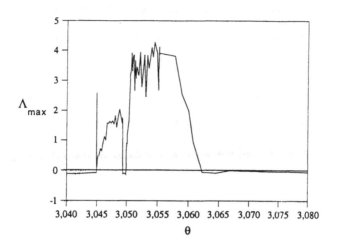

Figure 10.5. Lyapunov exponents for the case $\kappa = 4/450$, $\beta = 4/15$, $\lambda = 225$, $\sigma = 2$, $\rho = 20$.

exists in a finite band of the parameter, θ, values. The boundary of the regions of periodic and chaotic motions are summarized in Table 10.2. The first period-doubling

θ Region	Lyapunov Exponents	Attractor
$\theta < 2983.245$	$\Lambda < 0$	stable fixed point
$2983.245 < \theta < 3044.945$	$\Lambda = 0$	stable periodic solutions
$3044.945 < \theta < 3049.55$	$\Lambda > 0$	chaotic behavior
$3049.55 < \theta < 3050.0$	$\Lambda = 0$	stable periodic solutions
$3050.0 < \theta < 3060.5$	$\Lambda > 0$	chaotic behavior
$3061.0 < \theta < 7305.47$	$\Lambda = 0$	stable periodic solutions

Table 10.2. Lyapunov exponents for the case $\kappa = 4/450$, $\beta = 4/15$, $\lambda = 225$, $\sigma = 2$, $\rho = 20$.

bifurcation parameters accumulate at a critical value after which the model becomes chaotic. This value is found approximately to be 3044.945, which compares well to the limiting value obtained using Feigenbaum number. As the parameter is varied, Λ_{max} dips back below the axis as the periodic motions are encountered, in particular at $\theta \approx$ 3049.55. Periodic solutions in this regime again undergo secondary bifurcations, again leading to chaotic behaviour. As the bifurcation parameter θ is changed further, the Lyapunov exponent eventually passes through zero as the model undergoes periodic motions. The results of this technique will be checked in the following sections for sixteen values for θ.

10.2.2 Phase Trajectories

Inspection the system trajectories in the phase space is one of the methods to identify nonperiodic motions. The numerical integration of the model equations was performed using Gear's method for integration of stiff systems of ordinary differential equations. Most of the results were also checked by using the Runge-Kutta-Verner fifth-order and six-order methods. The time series and the projection of the trajectories in the x-y, x-z, y-z planes for values of θ between 3030 and 3062 are plotted in Figure 10.6.

The forms of the trajectories in the phase plane for the first six cases, (Figure 10.6a, b, c, d, e, f), give an intuitive feeling on the period-doubling bifurcation process specially for low-order period-doubling, say, period-32. In each case, the trajectory shows an orbit which crosses itself but it is still closed. In Figure 10.6g, h and i the chaotic oscillations have orbits which never close or repeat. The trajectory of the orbits tend to fill up a section of the phase space. The cases j, k, and l correspond to the region 3049.55 < θ < 3050.0 which is given in Table 10.2. We have found that case (j) contains a limit cycle of period-3. If a family of solutions has a 3-periodic orbit, it also

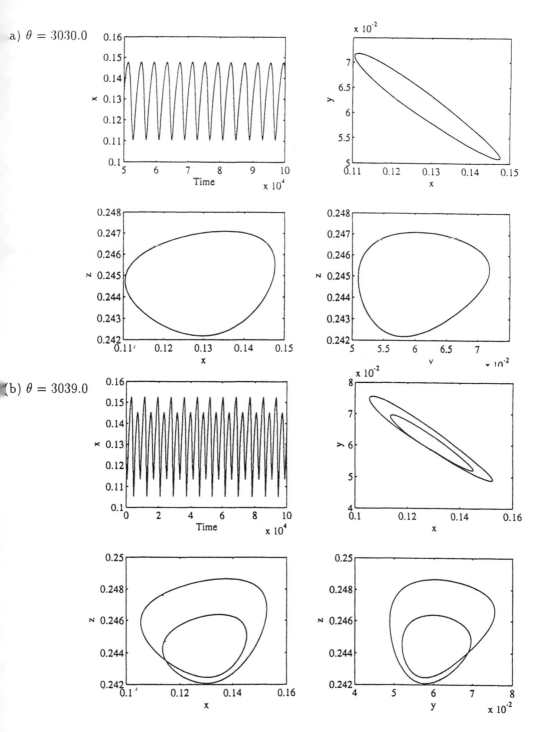

Figure 10.6. Time history and phase planes.

Figure 10.6 (continued)

(c) $\theta = 3043.5$

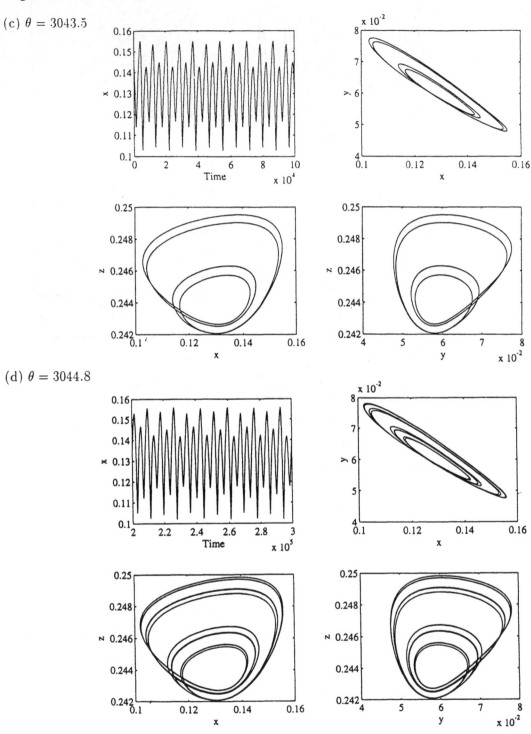

(d) $\theta = 3044.8$

Figure 10.6 (continued)

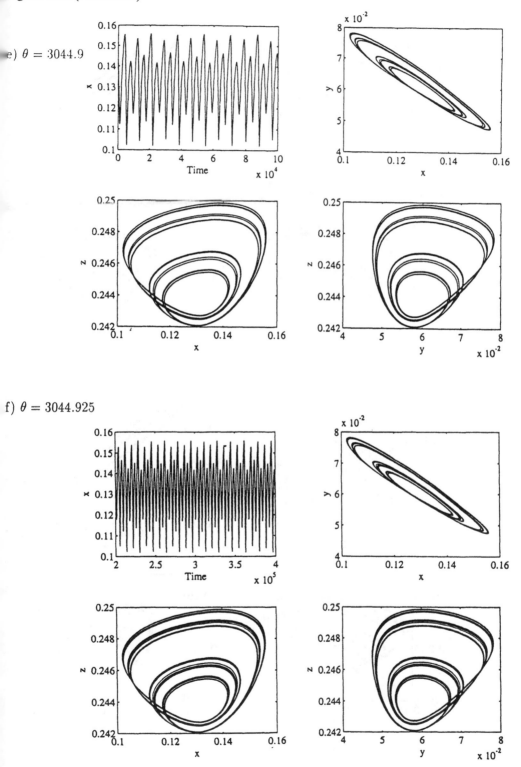

e) $\theta = 3044.9$

f) $\theta = 3044.925$

Figure 10.6 (continued)

(g) $\theta = 3045.0$

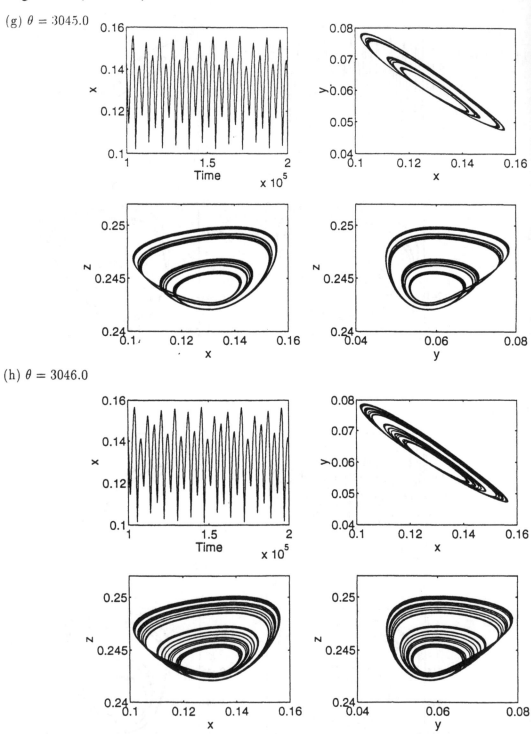

(h) $\theta = 3046.0$

Figure 10.6 (continued)

i) $\theta = 3049.0$

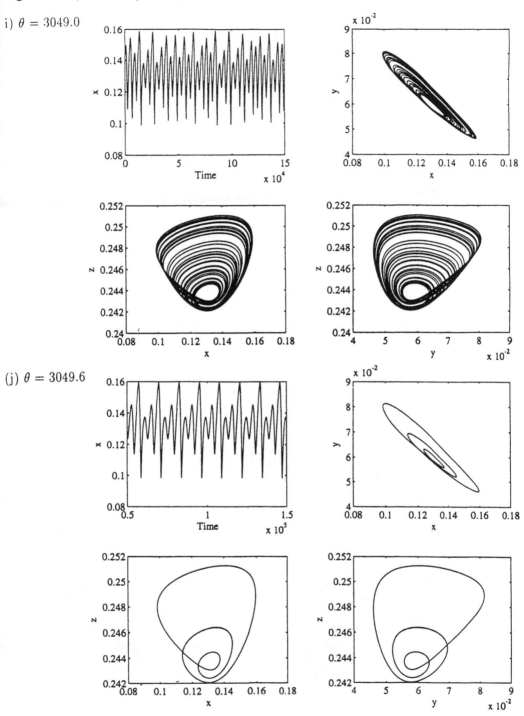

(j) $\theta = 3049.6$

Figure 10.6 (continued)

(k) $\theta = 3049.8$

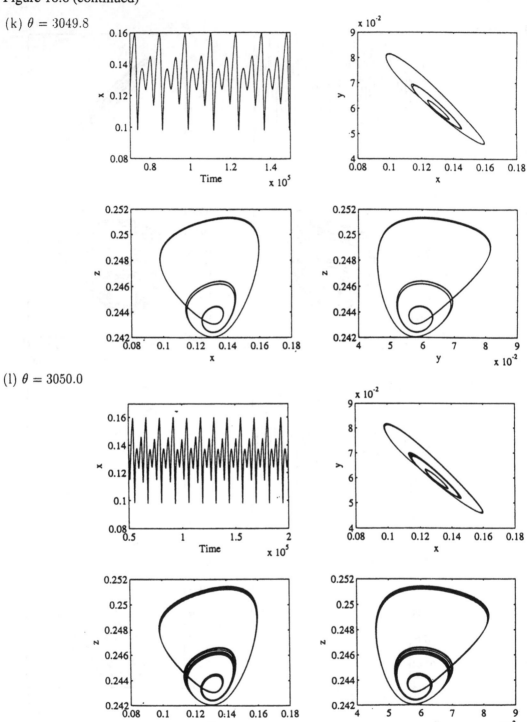

(l) $\theta = 3050.0$

Figure 10.6 (continued)

(m) $\theta = 3051.0$

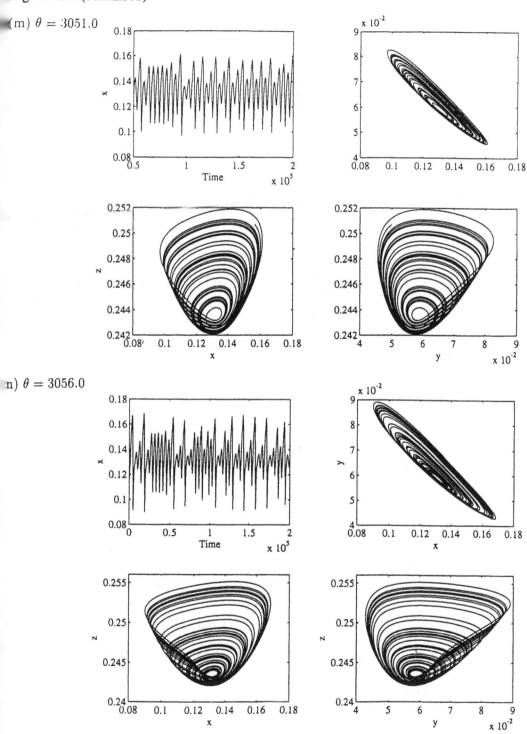

n) $\theta = 3056.0$

Figure 10.6 (continued)

(o) $\theta = 3060.0$

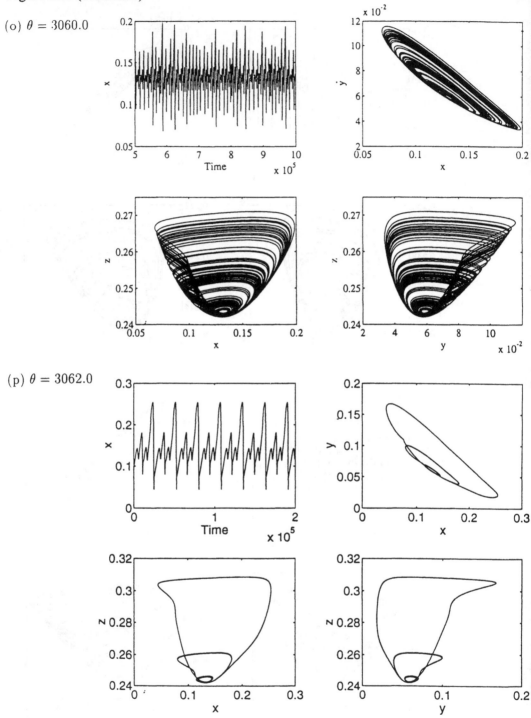

(p) $\theta = 3062.0$

has orbits of arbitrarily long period, according to the famous statement "period three implies chaos" of Li and Yorke. Figure 10.2k and l show period-6 limit cycles and large-period limit cycles. Therefore, periodic motions in this regime again undergo period-doubling, which again leads to chaotic behaviour. For the chaotic oscillations defined in the interval $3050.0 < \theta < 3060.5$, the phase plane trajectories for three cases are shown in Figure 10.6m, n, and o, respectively. For the last case, $\theta = 3062$, the model has a limit cycle of period-4 (Figure 10.6p). This suggests that the model exhibits period-halving leading to period one limit cycles.

The trajectories in the three-dimensional x - y - z space for all cases are shown in Figure 10.7. These 3-D curves are of interest, they delimit the shape of the attractor which the trajectories are following. It can be seen that the chaotic attractors can on divided into two parts: on the top part of the attractor which lies above the locus of the unstable equilibrium point, the neighboring trajectories are roughly parallel or diverging, while in the bottom part, the trajectories are moving close to each other. This is a property of chaotic attractors. **Figure 10.8** shows the state space trajectories during a chaotic motion at different times. It can be observed that these orbits never close or repeat. This wandering of orbits is a one of the manifestations of chaos.

259

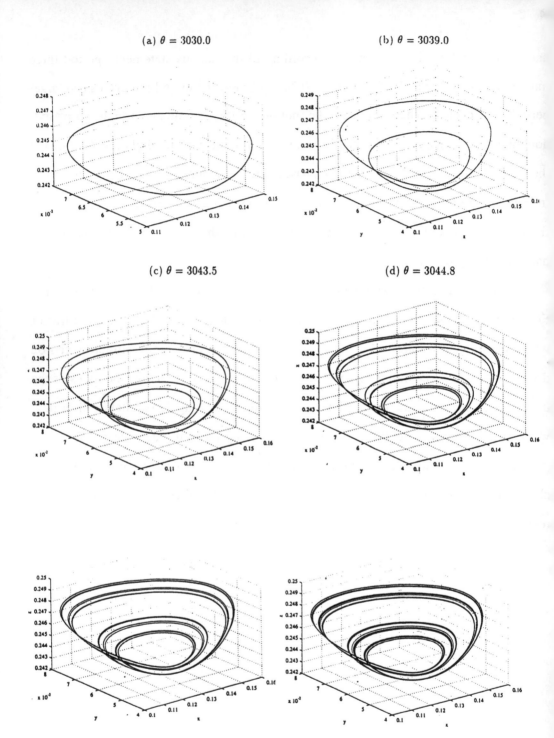

(a) $\theta = 3030.0$

(b) $\theta = 3039.0$

(c) $\theta = 3043.5$

(d) $\theta = 3044.8$

Figure 10.7. Trajectories in the x, y, z-space.

Figure 10.7 (continued)

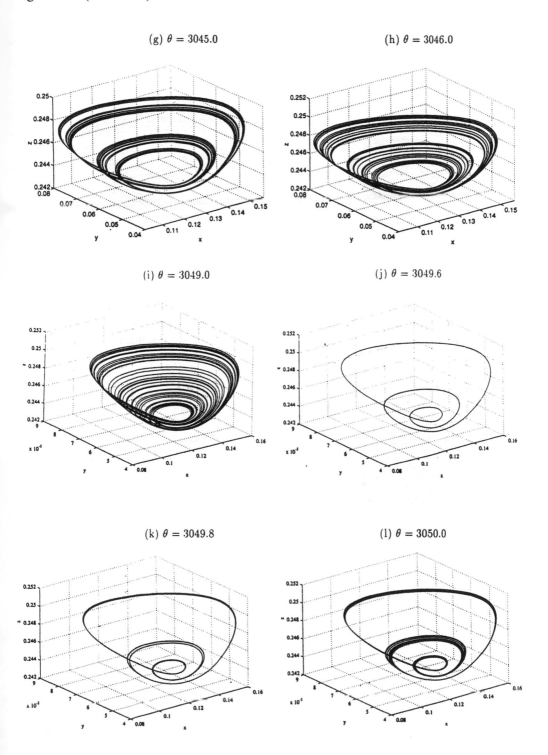

(g) $\theta = 3045.0$

(h) $\theta = 3046.0$

(i) $\theta = 3049.0$

(j) $\theta = 3049.6$

(k) $\theta = 3049.8$

(l) $\theta = 3050.0$

Figure 10.7 (continued)

(m) $\theta = 3051.0$ (n) $\theta = 3056.0$

(o) $\theta = 3060.0$ (p) $\theta = 3062.0$

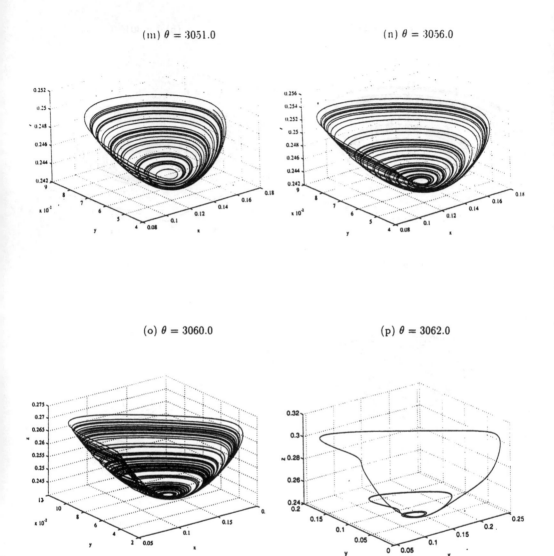

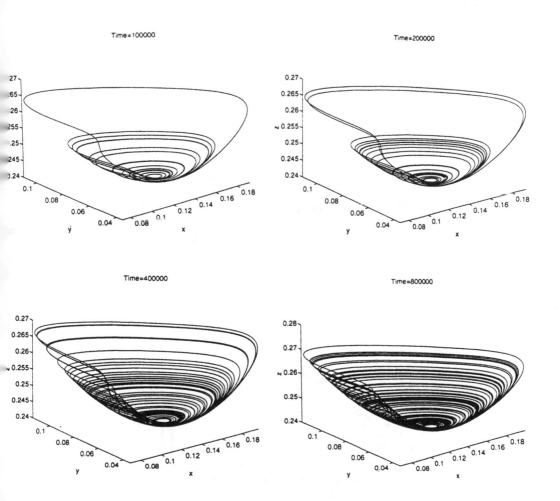

Figure 10.8. Chaotic attractor ($\theta = 3060$).

10.2.3 Poincaré Map

We now look at the sequence of points in phase space generated by the penetration of the continuous trajectory through a transverse surface or plane in the space. This technique, known as Poincaré map, will be used for studying all cases presented in last section. As a fact, the appearance of fractal-like structures in the the Poincaré Map is a strong indicator of chaotic motion. We examine the model trajectories when crossing the plane defined by

$$y = 0.06, \quad x \leq x_{ss}, \quad z \leq z_{ss}$$

where x_{ss} and z_{ss} are the values of the unstable equilibrium point. This plane is chosen that cuts the attractor transversally (Figure 10.9). This plane cuts part of the attractor on which neighboring trajectories are being folded back together. The Poincaré Map for the

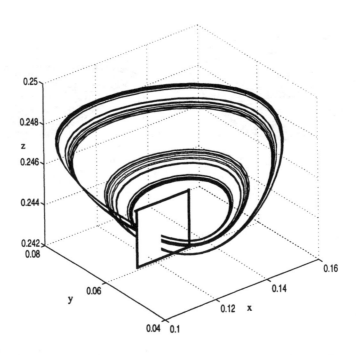

Figure 10.9. Poincaré map, a plane is chosen so that it cuts the trajectories transversally: $y = 0.06$.

mit cycle of period p shows p points in the plane. This is illustrated in Figure 10.10a, b, , d, e, f, j, k, l, p. The maps for the chaotic motions shown in Figure 10.10g, h, i, m, n, o ontain uncountable number of points which looks like an open curve.

Now we explore the fine structures 'fractals' embedded in the chaotic oscillations. Consider the chaotic motion at $\theta = 3060$. Figure 10.10o shows the result for plotting 2000 successive points. Figure 10.11a shows the result of 24000 points starting from different initial conditions. The two figures are seen to be almost identical. For both figures the transient regimes were eliminated. The resulting structure is known as a strange attractor. Trajectories originating points which lie off this structure move on to it, but do not then settle to any normal periodic motions. This structure, which is a fractal object, appears to consist of a number of parallel curves. The points tend to distribute themselves densely over these curves. In Figure 10.11b, we show a magnified view of a small region of Figure 10.11. It shows number of visible curves. One more enlargement results in Figure 10.11c. Here the points becomes more sparse, but parallel curves can still be traced.

We have seen that the mathematical results for one dimensional maps such as period-doubling and Feigenbaum number apply for our model. We will find some connection between our model and one-dimensional mapping by examining a special case of return map, the next-maximum map. The next-maximum or amplitude map plots the maximum of the $(n+1)$ peak against the maximum of the nth. Also it can be seen as no more than a particular case of the Poincaré map where the plane is defined by the condition

$$\frac{dx}{d\tau} = 0$$

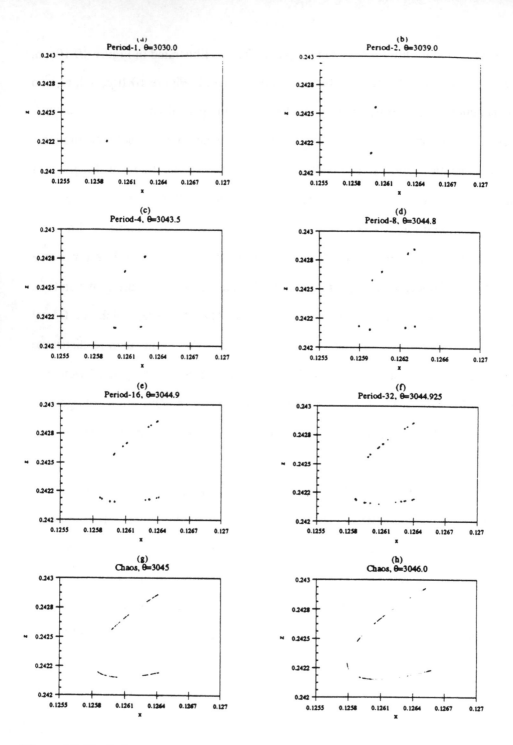

Figure 10.10. Poincaré maps.

266

Figure 10.10 (continued)

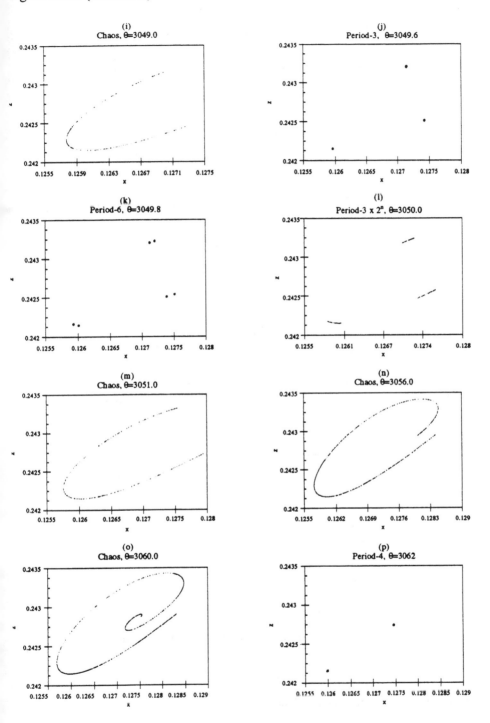

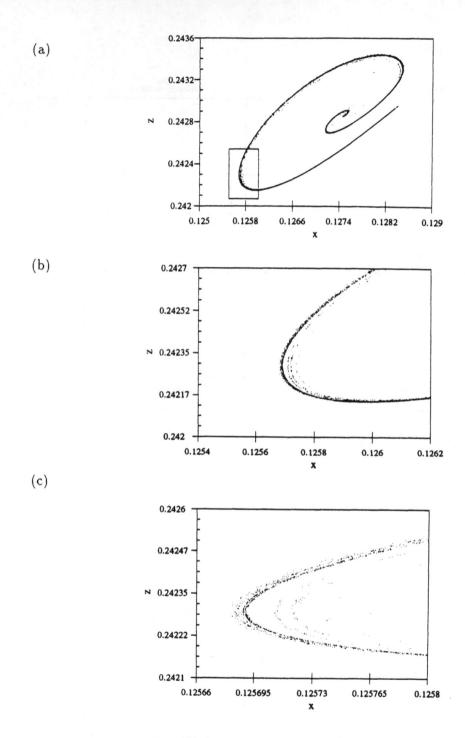

(a)

(b)

(c)

Figure 10.11. Poincaré map for the case $\theta = 3060$.

For regular periodic motions this map will consist of finite number of discrete points. Figure 10.12a shows the next-maximum plot for the case of period-16. There are only sixteen points which correspond to the sixteen peaks of the periodic solution. The order of these points is indicated by numbering 16 successive points in Figure 10.12a. For strange attractors this process will result in infinite number of different points which appear to lie on some open curve. If this curve appear to be clustered in some apparent functional relation and it has a maximum or a minimum, then the system can be declared to be chaotic. If this be the case, we can attempt to find a polynomial function that fit the data, $x_{n+1} = f(x_n)$, and use this mapping to do analysis. Renormalization theory [Moon, 1992; Scott, 1991] has shown that all maps with single, broad maxima will show basically the same sequences to chaos through period-doubling and give rise to the same value for the Feigenbaum number. For our model, Figure 10.12b, c, d, e, and f show the maps for chaotic traces for $\theta = 3046.0, 3049.0, 3051.0, 3056.0, 3060.0$. The first three cases show curves which satisfy the two criteria needed to declare the system chaotic. First, it can be approximated by a simple cubic polynomials and second, it has a relatively single broad maxima. In the chaotic region for $\theta = 3056.0, 3060.0$, the next-maximum maps have a sharp peak. This sharp peaks have been found with other models such as Lorenz model. The map around this peak is similar to a tent map defined in the region [0,1] as [Moon, 1992]:

$$x_{n+1} = rx_n, \quad x_n < 0.5$$

$$x_{n+1} = 1 + r - rx_n, \quad x_n \geq 0.5; \ r > 1$$

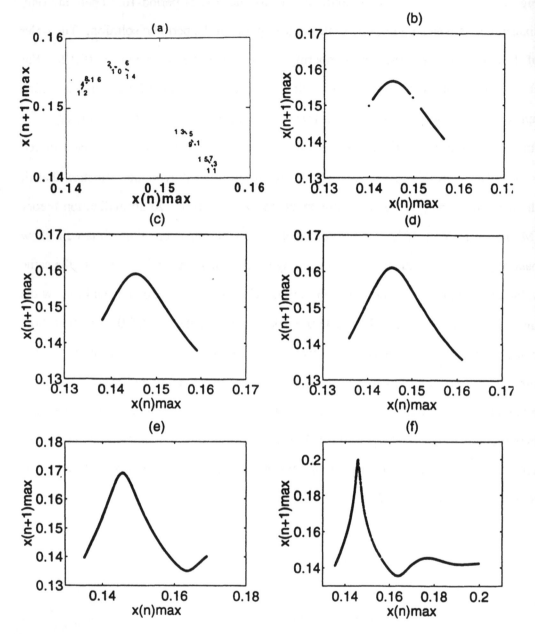

Figure 10.12. Next-maximum maps (a) $\theta = 3044.9$, (b) $\theta = 3046.0$, (c) $\theta = 3049.0$
(d) $\theta = 3051.0$, (e) $\theta = 3056.0$, (f) $\theta = 3060.0$.

270

This form of map can still support a period-doubling route to chaos but may have different scaling properties.

0.2.4 Power Spectra

One of the clues to chaos is the appearance of a broad spectrum of frequencies. Power spectra analysis in the frequency domain is obtained using 8192 points fast Fourier transform. Although using 8192 points power spectra analysis, we can hardly go beyond the 32-period motions, but it is still very useful in telling chaotic bands from periodic orbits.

In the regular periodic motions, the change from period-1 to period-2 shows the doubling of period of the oscillation. The Fourier transform shows halving of the original frequency $f/2$. In further period-doubling, say in period $2n$, not only will $f/2n$ be present, but equally likely one will see other peaks $mf/2n$. This is illustrated in Figure 10.13a, b, and c for period-1, period-2 and period-4 cases, respectively. Comparing the Fourier spectrum for the periodic motions (Figure 10.13a, b, c, d, f, j, k, l, o) and the chaotic behaviour (Figure 10.13g, h, i, m, n), we can see that the Fourier transform power spectrum for the chaotic states show much broader band structure than those for periodic motions.

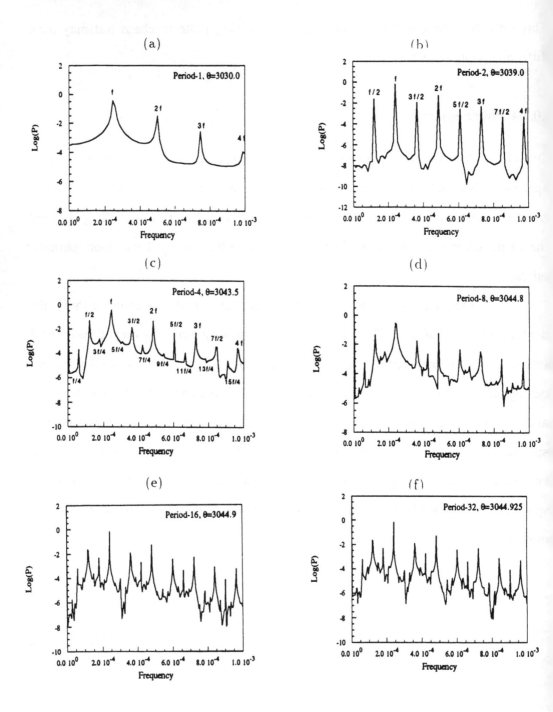

Figure 10.13. Power spectra

272

Figure 10.13 (continued)

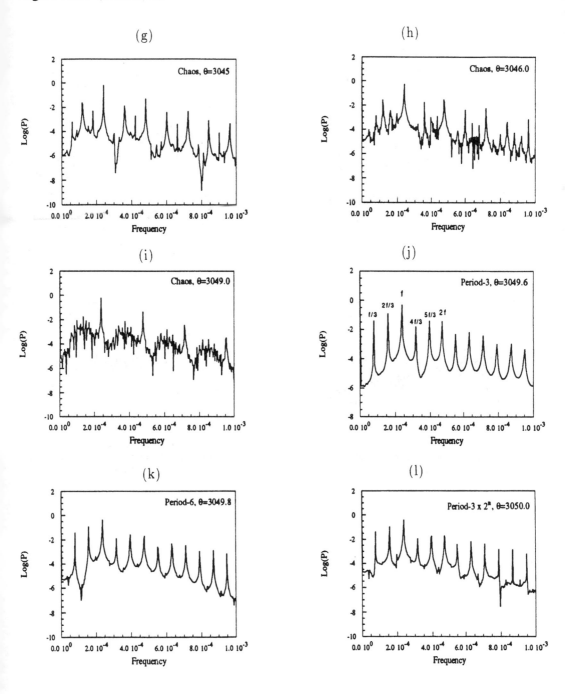

Figure 10.13 (continued)

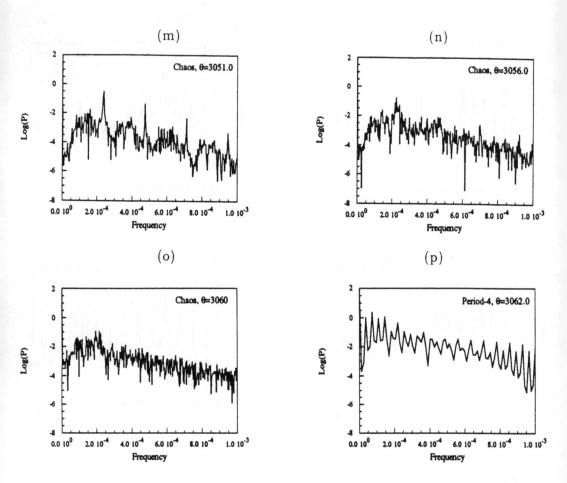

Chapter 11

CHAOS - MODEL II

The mathematical model for the second case is

$$\frac{dx}{d\tau} = \frac{1-x}{\theta} - xy^2$$

$$\frac{dy}{d\tau} = \frac{z-y}{\lambda} - \frac{y}{\theta} + xy^2 - \kappa y$$

$$\rho\frac{dz}{d\tau} = -\frac{z-y}{\lambda} + \frac{\sigma}{\theta}(\beta - z) - \rho\kappa z \qquad (11.1)$$

For this model, we have observed sequences of period-doubling in distinct regions of the parameter space. In the first section, we define a region in the parameter space κ, β where the system exhibits period-doubling bifurcations. In the second section, we show the transition to chaos via period-doubling bifurcation which gives rise to Feigenbaum's number. Diagnostic techniques such as Lyapunov exponents, Poincaré map and power spectra are used to examine these chaotic motions. In the third section we present another example of a chaotic attractor.

11.1   Period-Doubling in the Parameter Space

At present, there are no techniques available for a prior determination of regions of parameter values for which chaos will be possible. Period-doubling, one of the mechanisms of losing stability for periodic solutions, often occurs and leads to chaotic behaviour. Examining the period-doubling singularities in the parameter space is a good first move.

The parameter space for the case $\sigma = 0.1, \lambda = 100.0, \rho = 2.0$ is shown in Figure 11.1. It is found that the periodic solutions undergo period-doubling bifurcations for cases lying under the period-doubling boundary (PD). The PD curve is obtained by examining the mechanisms of the appearance/disappearance of period-doubling in the bifurcation diagrams. The first mechanism is illustrated in Figure 11.2a, b where two period-doubling bifurcations appear in the bifurcation diagram. This mechanism is analogous to the $\mathbf{H_{01}}$ degeneracy of Hopf points. Figure 11.2c, d shows that one of the period-doubling points disappear when it collides with the saddle equilibrium point. The second mechanism is the period-doubling-turning point interactions. Here the periodic solutions loses its stability via saddle-node bifurcation before period-doubling takes place. This is illustrated in Figure 11.2e and f. This mechanism takes place close to the $\mathbf{H_{10}}$ degeneracies where the periodic solutions change their stability at the Hopf point itself.

Now we look at the effect of the parameters σ, λ, and ρ on the period-doubling singularities. Consider the system parameter values

$$\kappa = 0.01, \beta = 0.2, \sigma = 0.1, \lambda = 100.0, \rho = 2$$

For these values, the system shows an isola pattern with one Hopf point. The periodic solution branch emanating from this Hopf point exhibits period-doubling bifurcation. We fix four of these parameters and vary the fifth parameter to examine the locations of the period-doubling bifurcations in the parameter space defined by this chosen parameter and the bifurcation parameter θ. Figure 11.3 shows the Hopf points curve and the

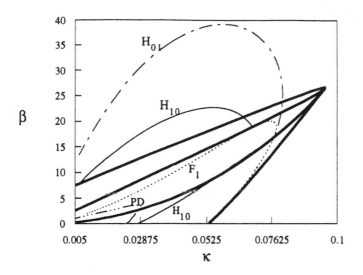

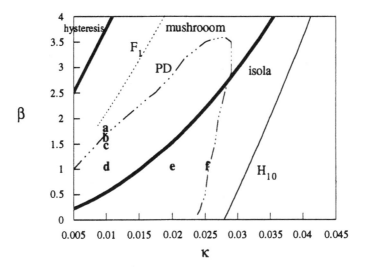

Figure 11.1. Parameter space ($\sigma = 0.1$, $\lambda = 100.0$, $\rho = 2.0$).

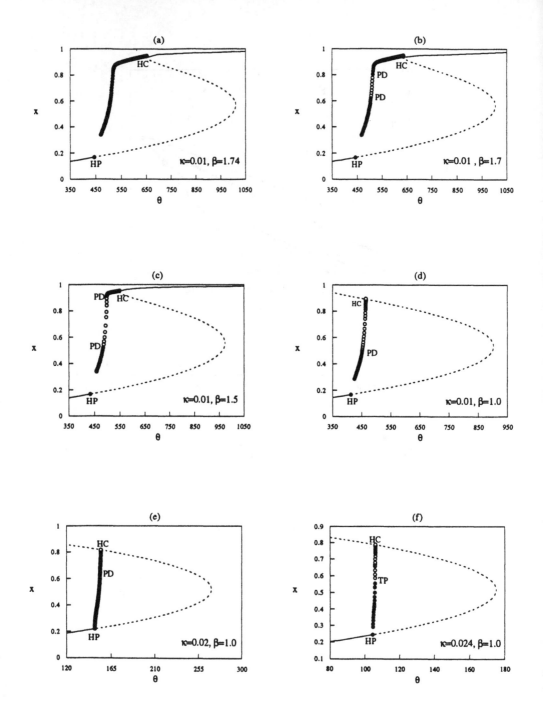

Figure 11.2. Mechanisms of the appearance of period-doubling bifurcations.

278

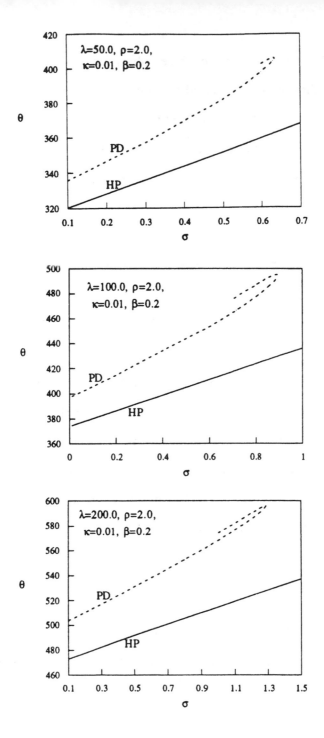

Figure 11.3. Period-doubling loci in the θ, σ-plane - effect of λ.

period-doubling points the θ, σ plane for three values for the mass-transfer resistance coefficient λ. In all these cases, the period-doubling loci have a maximum value for σ. Above these values, the model does not exhibit period-doubling bifurcation. Figure 11.3 shows also that increasing λ increases this maximum value of σ. In Figure 11.4, the effect of the tank-volume ratio on the values of σ_{max} is illustrated. It shows that increasing ρ also increases σ_{max}.

In general, period-doubling does not imply chaos, we show this by examining the case

$$\kappa = 0.01, \beta = 0.2, \sigma = 0.8, \lambda = 100.0, \rho = 2$$

For these values, Figure 11.5 shows that the period-doubling bifurcation does not give rise to chaotic behaviour. The periodic solution undergoes a period-doubling bifurcation at $\theta = 476.58$, after which the system has period-2 oscillations. Increasing the bifurcation parameter θ further does not lead to period-4 but at $\theta = 487.901$ it returns the model back to period-one oscillations. The time series for five cases taken from the period-doubling range are shown in Figure 11.6.

11.2 Chaotic Behaviour - Case I

For low values of σ in Figures 11.3 and 11.4, we have observed a cascade of period-doubling bifurcations which are likely to be infinite. An example of such a behaviour is defined by the parameter values

$$\kappa = 0.01, \beta = 0.2, \sigma = 0.1, \lambda = 100.0, \rho = 2$$

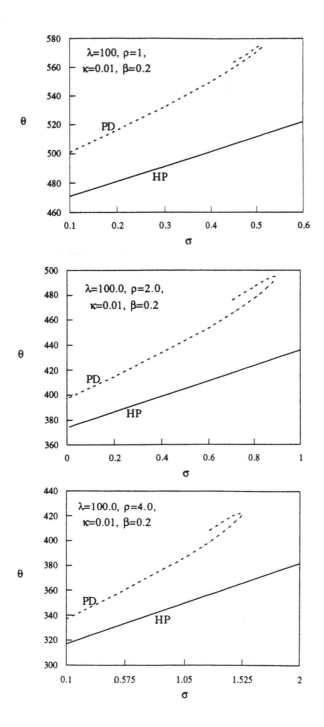

Figure 11.4. Period-doubling loci in the θ, σ-plane - effect of ρ..

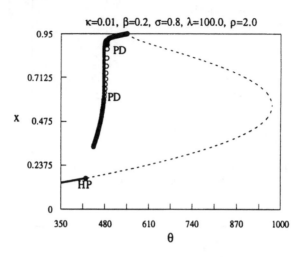

Figure 11.5. Bifurcation diagram ($\kappa = 0.01$, $\beta = 0.2$, $\sigma = 0.8$, $\lambda = 100.0$, $\rho = 2.0$.

For these values, we have:

- For $4.6338 < \theta \leq 380.2087$ the model exhibits only multiple steady states

- For $380.2087 < \theta \leq 418.6317$ the system gives rise to stable periodic solutions which loses its stability via period doubling bifurcation before the end of oscillatory branch.

- For $418.6317 < \theta \leq 798.284$ the model shows only multiple steady states.

Figure 11.7 shows that the period-doubling takes place at $\theta = 405.8076$. This is followed by a series of period-doubling bifurcations (Figure 11.7). Figure 11.8a shows a limit cycle in the phase plane for period-1, followed by Figure 11.8b, c, and d by limit cycles with period-2, period-4, and period-8.

282

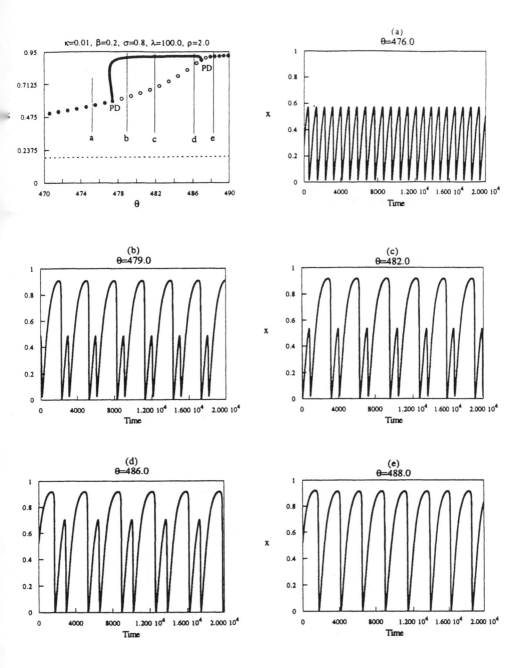

Figure 11.6. Time series for the case of finite period-doubling.

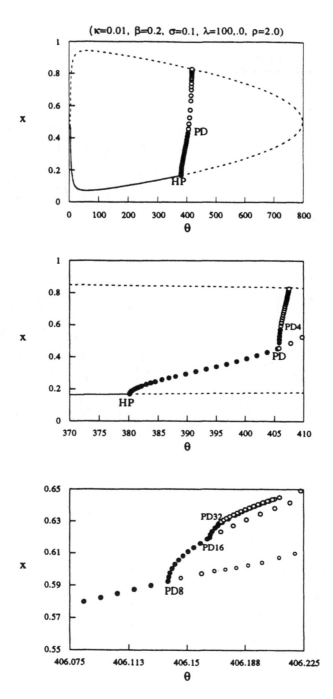

Figure 11.7. Bifurcation diagram ($\kappa = 0.01$, $\beta = 0.2$, $\sigma = 0.1$, $\lambda = 100.0$, $\rho = 2.0$).

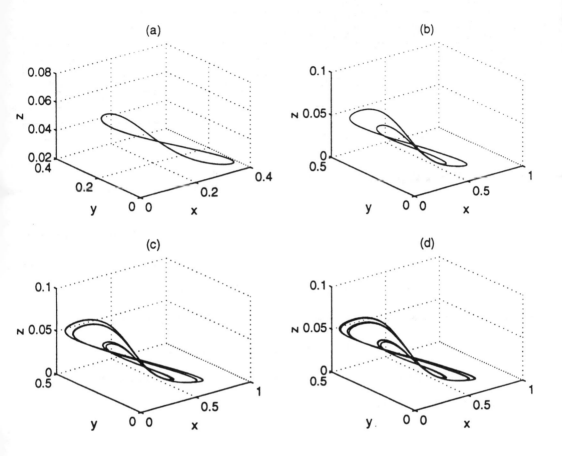

Figure 11.8. Limit cycles (a) $\theta = 400.0$, (b) $\theta = 406.0$, (c) $\theta = 406.164$, (d) $\theta = 406.168$.

11.2.1 Feigenbaum's Number

We have found that the period-doubling bifurcation cascade shown above gives rise to Feigenbaum's number. The first column in Table 11.1 shows the calculated values for six

Bifurcation Type	θ	δ	θ^*
Hopf	380.2077704289		
Period-2	405.8077612946		
Period-4	406.1378639945		
Period-8	406.1646089695	12.3457	
Period-16	406.1697417831	5.2105	
Period-32	406.1708016391	4.842935	406.1708410746
Period-64	406.17102733726	4.6959	406.1710765091762
Period-128		4.6692016	406.1711269320528
...	...	...	...
θ_∞		4.6692016	406.171140674271

Table 11.1. Scaling property of period-doubling bifurcations.

consecutive period-doubling bifurcation parameters. These values are used to estimate the Feigenbaum's number in the second column which is approximately equal to the proposed value 4.6692016. Therefore, this model is structurally equivalent to the difference equations with quadratic transformations which were examined by Feignbaum. In the fourth column of Table 11.1, the limiting value at which the chaotic dynamics develop is obtained by extrapolating the results. The accumulation point is found approximately to be 406.1711406739819.

11.2.2 Lyapunov Exponents

The calculations of the Lyapunov exponents for the region $406.16 < \theta < 406.2$ produce the results shown in Figure 11.9. It is found that the model displays quite interesting and complicated transitions between periodicity and chaos. The initial transition from periodic to chaotic behaviour occurs at $\theta = 406.171173$. This value compares well to the limiting value of θ shown in Table 11.1. There are also four transitions which produce narrow windows of periodic behaviour embedded within the major region of chaos ($406.171173 < \theta < 406.19366$ approximately). They are listed in Table 11.2.

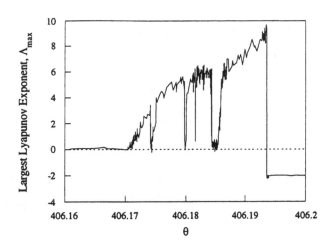

Figure 11.9. Dependence of the largest Lyapunov exponent on θ ($\kappa = 0.01$, $\beta = 0.2$, $\sigma = 0.1$, $\lambda = 100.0$, $\rho = 2.0$)

θ Region	Lyapunov exponents	Attractor
$380.2077 < \theta < 406.171173$	$\Lambda = 0$	periodic solutions
$406.171173 < \theta < 406.1741$	$\Lambda > 0$	chaotic solutions
$406.1741 < \theta < 406.1746$	$\Lambda = 0$	periodic solutions
$406.1746 < \theta < 406.17977$	$\Lambda > 0$	chaotic solutions
$406.17977 < \theta < 406.1802$	$\Lambda = 0$	periodic solutions
$406.1802 < \theta < 406.181615$	$\Lambda > 0$	chaotic solutions
$406.181615 < \theta < 406.18163$	$\Lambda = 0$	periodic solutions
$406.18163 < \theta < 406.18435$	$\Lambda > 0$	chaotic solutions
$406.18435 < \theta < 406.1856$	$\Lambda = 0$	periodic solutions
$406.1856 < \theta < 406.193634$	$\Lambda > 0$	chaotic solutions
$406.19364 < \theta$	$\Lambda < 0$	steady state

Table 11.2. Lyapunov exponents for the case ($\kappa = 0.01$, $\beta = 0.2$, $\lambda = 100.0$, $\sigma = 0.1$, $\rho = 2.0$)

11.2.3 Phase Trajectories and Power Spectra

In Figure 11.10 to 11.19 we examine the model trajectories for ten values of θ taken from the chaotic band (406.171173 < θ < 406.19366). In all these cases, the results of initial time units are discarded. Figure 11.10 shows that the model displays chaotic behaviour for θ = 406.174. As θ is increased to the value θ = 406.1743, the model shows regular periodic oscillation of period-12 (Figure 11.11). This value of θ belongs to the first transition region. Increasing θ further to 406.1775 brings the model back to chaotic behaviour as shown in Figure 11.12. In the second transition window, simulating the model around θ = 406.1798 gives period-10 oscillations displayed in Figure 11.13. Figure 11.14 shows the chaotic behaviour of the model for θ = 406.181 which is just after the transition to chaos. By increasing θ slightly to 406.181625, the model returns back to the periodic behaviour. For this case, Figure 11.15 shows periodic motion of period-14. Figure 11.16 shows that as we increase θ to 406.183, the transition back to chaotic behaviour occurs. For θ = 406.185 which belongs to the largest periodic window in the chaotic band (406.18435 < θ < 406.1856), the simulation of the model shown in Figure 11.17 gives period-six oscillations. In the ninth case shown in Figure 11.18, θ = 406.19, the model behaviour changes back to chaos. The Fourier spectrum for the periodic motions and the chaotic motions are also shown for the previous nine cases. It can be seen that for chaotic cases, the power spectrum show much broader band structure than for periodic motions. As θ is further increased, the chaotic attractor retains its shape, with the main change that the gaps between the upper and lower trajectories become more populated. This is illustrated in Figure 11.19 for the case θ = 406.1936.

The transitions from periodic to chaotic behaviour in the periodic windows are found to occur by non-period-doubling processes. They are observed by examining the time series for values of θ very close to the transition boundaries. First, we consider the

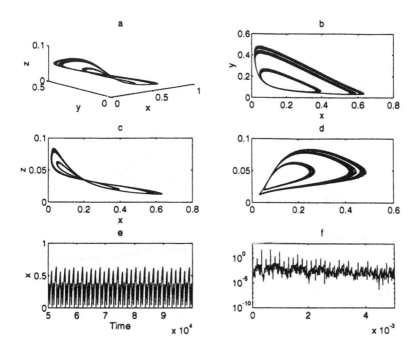

Figure 11.10. Model trajectories, $\theta = 406.174$: chaos

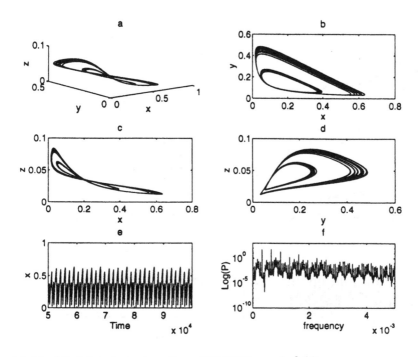

Figure 11.11. Model trajectories, $\theta = 406.1743$: period-12

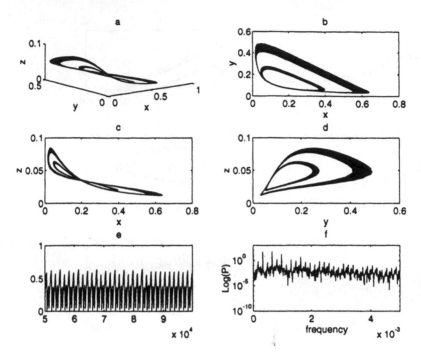

Figure 11.12. Model trajectories, $\theta = 406.1775$: chaos

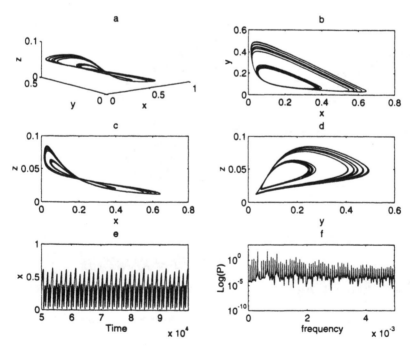

Figure 11.13. Model trajectories, $\theta = 406.1798$: period-10

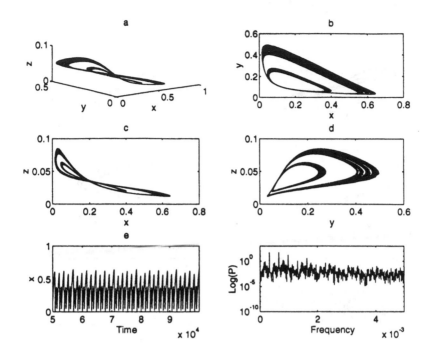

Figure 11.14. Model trajectories, $\theta = 406.181$: chaos

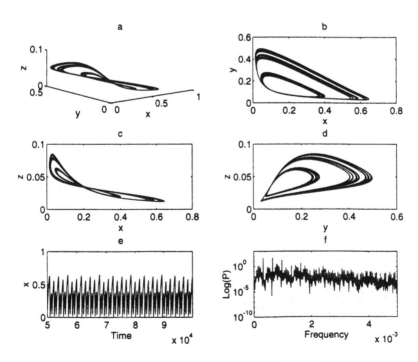

Figure 11.15. Model trajectories, $\theta = 406.181625$: period-14

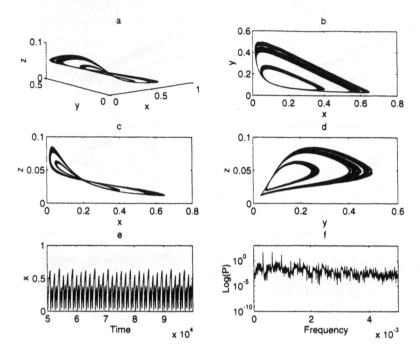

Figure 11.16. Model trajectories, $\theta = 406.183$: chaos

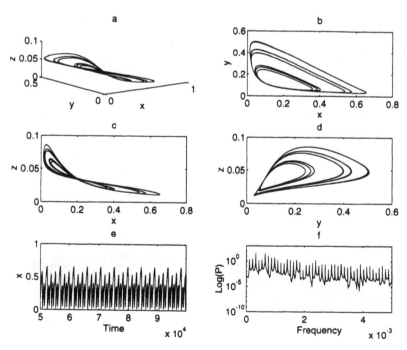

Figure 11.17. Model trajectories, $\theta = 406.185$: period-6

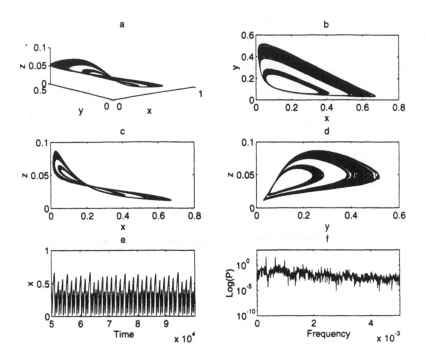

Figure 11.18. Model trajectories, $\theta = 406.19$: chaos

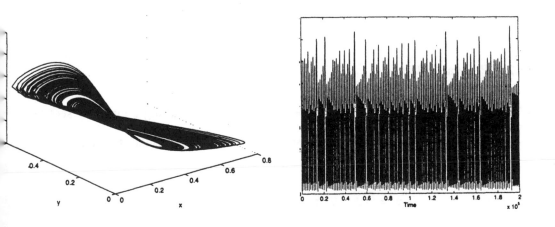

Figure 11.19. Model trajectories, $\theta = 406.1936$: chaos

293

first transition region ($406.1741 < \theta < 406.1746$). The time series for $\theta = 406.17408$ and $\theta = 406.1741$ are shown in Figure 11.20a,b, respectively. Integrating the model around $\theta = 406.1741$ gives regular oscillations of period-12 for random choice of initial conditions, while for $\theta = 406.17408$, these oscillations are interrupted by bursts (compare these two cases around $\tau \times 10^{-5} = 5, 7, 8.5, 9.7$). The time series of x for $\theta = 406.1745$ and 406.1746 which are at the other side of this transition region, are shown in Figures 11.21a, b, respectively. It is seen that the periodic oscillation is intermittently broken by portions of nonperiodic behaviour (e.g., around $\tau \times 10^{-5} = 5, 5.7, 6.2, 6.5, 7.8, 9.2$). In both sides of the second window $406.17977 < \theta < 406.1802$, the transition processes involve intermittency. These transitions are illustrated in Figure 11.22 and 11.23. In the largest periodic break, $406.18435 < \theta < 406.1856$, the chaotic transitions occur also via intermittency phenomena. In Figure 11.24a, b we show the time dependence of x for $\theta = 406.18435$ and 406.1844, respectively. It is seen that the periodic behaviour (period-6) in Figure 11.24a is interrupted by bursts around $\tau \times 10^{-5} = 0.6, 1.8, 3.5, 4.0$ approximately. In the other side of this window, the transition back to chaotic behaviour occurs also by intermittency route (Figure 11.25a, b).

11.2.4 Poincaré Map

We now study the structure of the chaotic attractor by examining a Poincaré section of the attractor. Such a transverse plane is shown in Figure 11.26 which is defined by

$$y = 0.1, \quad 0 < x < 0.15, \quad .04 < z < 0.06$$

This plane cuts part of the attractor on which the neighboring trajectories are moving close to each other. For $\theta = 406.19$, the results of ten thousands points of intersections

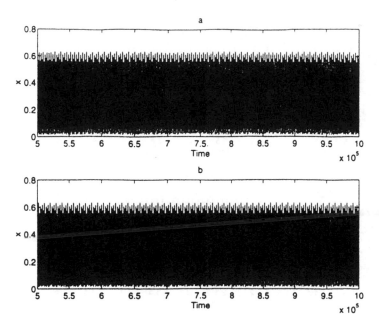

Figure 11.20. Intermittency route to chaos (a) $\theta = 406.17408$, (b) $\theta = 406.1741$.

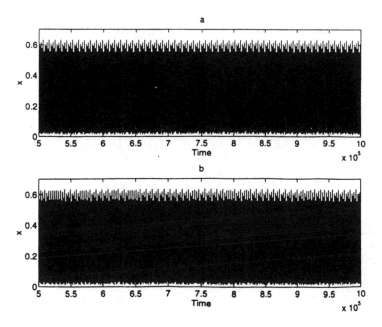

Figure 11.21. Intermittency route to chaos (a) $\theta = 406.1745$, (b) $\theta = 406.1746$.

295

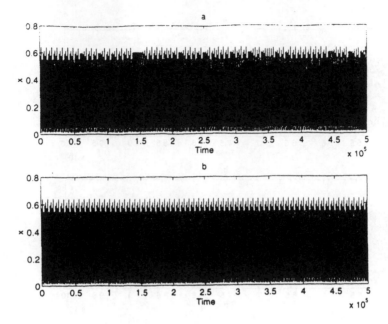

Figure 11.22. Intermittency route to chaos (a) $\theta = 406.7975$, (b) $\theta = 406.17977$.

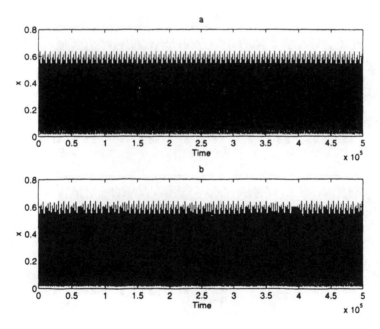

Figure 11.23. Intermittency route to chaos (a) $\theta = 406.18$, (b) $\theta = 406.1802$.

296

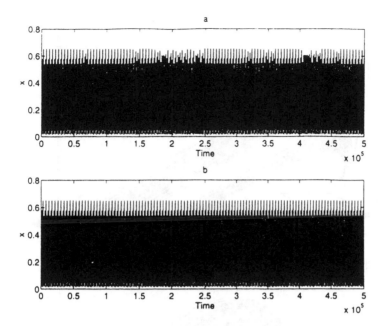

Figure 11.24. Intermittency route to chaos (a) $\theta = 406.18435$, (b) $\theta = 406.1844$.

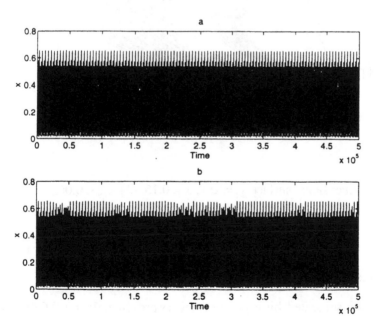

Figure 11.25. Intermittency route to chaos (a) $\theta = 406.1858$, (b) $\theta = 406.186$.

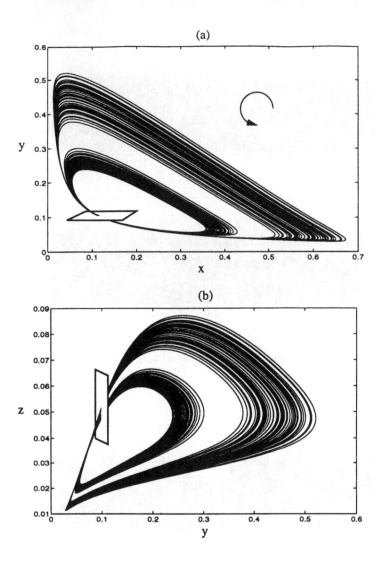

Figure 11.26. Two-dimensional surface $y = 0.1$, $x < 0.15$, $.04 < z < 0.06$.

are shown in Figure 11.27a. These points tend to distribute themselves over two smooth one-dimensional curves with a maximum of $x = .126535$ and a minimum of $x = .12095$. These two curves are separated by a gap defined approximately by $0.0498 < z < 0.0509$. As θ is further increased to 406.1936 these two curves becomes elongated and hifted to higher values of x and the gap between these two curves has been also reduced Figure 11.27b).

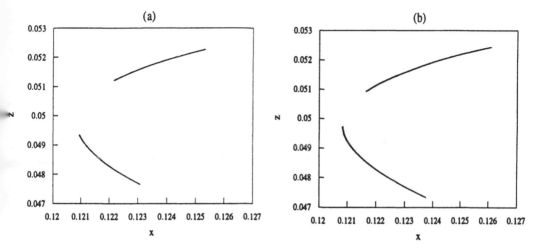

Figure 11.27. Poincaré map (a) $\theta = 406.19$, (b) $\theta = 406.1936$.

The intersections of eight successive points for the case $\theta = 406.19$ are shown in Figure 11.28. It can be seen that each curve can be established using a smaller transverse section. The top curve (a) in Figure 11.28 can be obtained by using the plane

$$y = 0.1, \quad x < 0.15, \quad .05 < z < 0.06$$

while the other curve (b) can be found by the plane

$$y = 0.1, \quad x < 0.15, \quad .04 < z < 0.05$$

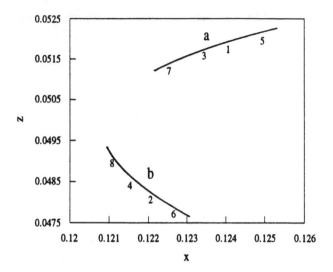

Figure 11.28. Poincaré map $\theta = 406.19$.

The return maps for both curves are shown in Figure 11.29. The return map is obtained
by plotting each x_n against its successor value x_{n+1}. For both cases, the return map points
become spread out along a single curve. These curves meet two criteria that have been

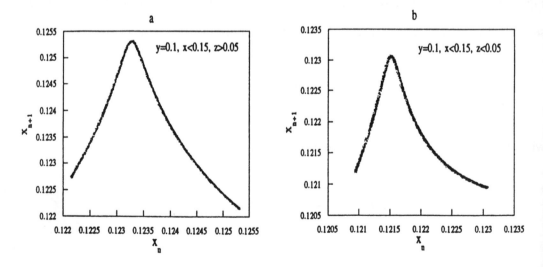

Figure 11.29. Return map.

used to declare the model chaotic; first, they have a maximum, and second, each curve can be fitted to a polynomial function, $x_{n+1} = f(x_n)$.

11.3 Chaotic Behaviour - Case II

In Section 8.2, we have seen that the model gives rise to period-doubling bifurcation for cases taken along the F_2 degeneracies. For example, for $\kappa = 0.00445$, $\beta = 1.0$, $\sigma = 0.1$, $\lambda = 20.0$, $\rho = 20.0$, Figure 8.17 shows that the model undergoes period-doubling bifurcation at $\theta = 153.3303204208$. In the following, we will show that chaos can exist in the vicinity of the F_2 and G_1 singularities.

11.3.1 Feigenbaum's Number

The values at which successive period-doubling bifurcations take place are summarized in Table 11.3. It shows that the ratio of the widths of successive θ intervals approaches Feigenbaum's number of 4.6692. Extrapolating the results in Table 11.3 indicates that infinite-period bifurcation occurs approximately at $\theta_\infty = 153.448153622475$!

Bifurcation Type	θ	δ	θ^*
Hopf	142.2605871528		
Period-2	153.3303204208		
Period-4	153.4251474174		
Period-8	153.4432482648	5.23882	
Period-16	153.4471014109	4.69768	
Period-32	153.4479282709	4.659974	
Period-64	153.4480972753		
...	...	...	...
θ_∞		4.6692016	153.448153622475

Table 11.3. Scaling property of period-doubling bifurcations.

11.3.2 Lyapunov Exponents

The calculation of Lyapunov exponents for this case produces the results shown in Figure 11.30. It shows that the model transforms into chaotic responses approximately at $\theta = 153.44825$. It can be seen also that the model exhibits chaos in the region $153.4482 < \theta < 153.4782$.

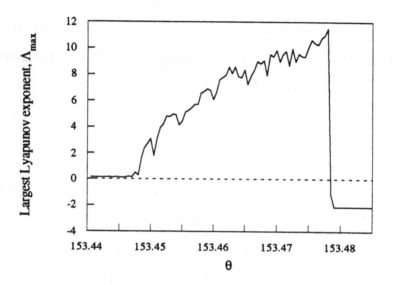

Figure 11.30. Lyapunov exponents ($\kappa = 0.00445$, $\beta = 1.0$, $\sigma = 0.1$, $\lambda = 20.0$, $\rho = 20.0$).

11.3.3 Phase Trajectories

Figure 11.31 shows four representative three-dimensional attractors for increasing values of θ. Figure 11.31a, b, and c show limit cycles of period-one, period-two and period-four, respectively. As θ is increased past the accumulation point, the model shows chaotic behaviour which is displayed in Figure 11.31d.

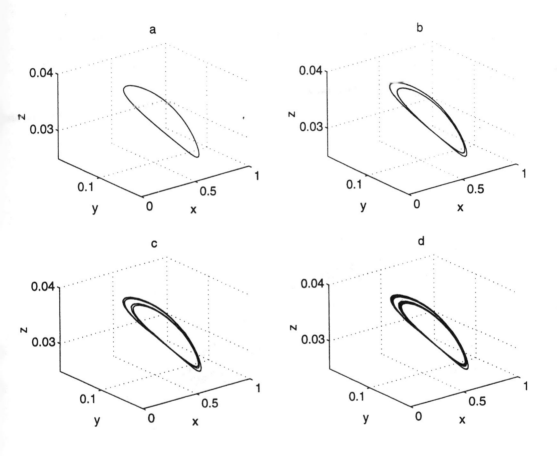

Figure 11.31. Trajectories in the x, y, z-space: (a) $\theta = 153.3$, (b) $\theta = 153.4$, (c) $\theta = 153.44$, (d) $\theta = 153.45$.

11.3.4 Poincaré Map

We now show the model trajectories when crossing the plane defined by

$$y = 0.08, \quad 0.525 < x < 0.575, \, 0.029 < z < 0.03$$

This transverse plane is shown in Figure 11.32. For $\theta = 153.475$ the results of several thousand intersections are shown in Figure 11.33. It shows a *fractal-like* structure which is a strong indicator of chaotic behaviour.

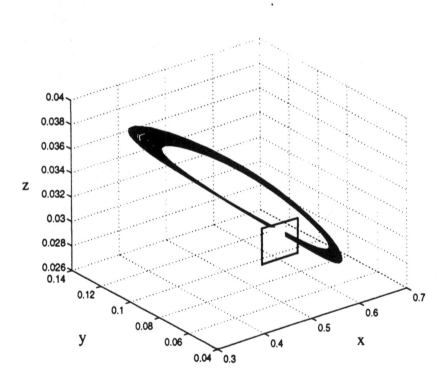

Figure 11.32. Model trajectories $\theta = 153.475$.

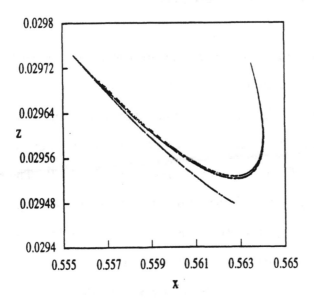

Figure 11.33. Poincaré map.

Chapter 12

SUMMARY AND CONCLUSIONS

We have studied the dynamics of a two-compartment (reservoir-reactor) chemical system, based on cubic autocatalysis; $A + 2B \rightarrow 3B, B \rightarrow C$. The coupling between the two compartments takes place by mass transfer of either the reactant A (Model I) or the autocatalyst B (Model II). The autocatalytic reaction occurs only in the reactor, but the degenerative reaction $B \rightarrow C$ takes place in the reservoir in Model II. The nonlinearity source is created by the presence of the third order term in the rate expression of the autocatalytic reaction ($r = k_1 AB^2$). This system is considered to be a simple extension of Gray and Scott's model of cubic autocatalysis in a CSTR.

The mathematical form of each model consists of three ordinary differential equations with six parameters; θ, residence time, κ, decay rate constant, β inflow autocatalyst concentration, σ, flow-rate ratio, λ, mass transfer resistance, and ρ volume ratio. The main tools in our analysis are the bifurcation diagrams of x vs θ; and the branch set in κ, β-parameter space. We have applied the local bifurcation methods (singularity theory, Lyapunov-Schmidt reduction technique and normal form method) to divide the branch set into regions of different static and dynamic patterns. We have also used the Doedel's program AUTO [Doedel and Kerevez, 1981] to generate the bifurcation diagrams of this system. In the following sections we summarize the results of our work, drawing attention to the differences from the behaviour of the Gray/Scott reaction in a CSTR.

12.1 Steady-state Analysis

• The steady-state conditions for both models are expressed in terms of a single polynomial equation, cubic in x_{ss}. Therefore, the models have either one or three steady states.

• The steady-state equations for Model I depend on all parameters except the volume ratio parameter ρ, while the steady-state conditions of the second model depend on all parameters. Thus, the parameter ρ does not have any influence when there is no reaction in the reservoir.

• Singularity theory has been successfully applied to these coupled models. The steady-state results for both models have shown that five different types of equilibrium patterns can be found in the κ, β-space for fixed σ, λ, and ρ. Three patterns have been seen in the case of the Gray and Scott CSTR: unique, isola, and mushroom. A fourth pattern is that of a single hysteresis loop. For the second model the hysteresis loop has the shape of the figure '2' where as for the first model, it is reversed and has the shape of an 'S'. The fifth pattern of response is found by unfolding the pitchfork points which are the highest singularity found for these models. Here, both hysteresis and an isola can be obtained over a narrow range of parameter values.

• It has been shown that winged cusp singularity does not occur for any of these models, even though the equilibrium patterns of the first model, Gray and Scott CSTR and second model are equivalent to the perturbed winged cusp singularity with $\alpha_3 < 0$, $\alpha_3 = 0$, and $\alpha_3 > 0$, respectively (α_3 is the second parameter in the universal unfolding of the winged cusp singularity, see Table 3.3 and Figure 3.4).

- When the mass transfer resistance vanishes, $\lambda \rightarrow 0$, both models show a multiplicity behaviour in the parameter space that is analogous to the Gray and Scott CSTR.

- The branch set of the second model (coupling through the autocatalyst) shows a greater sensitivity to the variation of the parameters: flow rate ratio, σ; mass transfer resistance, λ; and volume ratio, ρ.

- The effects of the parameters σ and λ on the branch set of the first model and of the parameters σ, λ, and ρ on the branch set of the second model are studied. It has been shown in the first model that increasing σ or decreasing λ enlarges the multiplicity region in the κ, β-space, while the opposite effect has been observed in the second model.

12.2 Dynamic Analysis

- The degenerate Hopf bifurcations arise when one or more of the Hopf Theorem conditions break down. We have made use of singularity theory and normal form method to examine the dynamic behaviour of these models. The dynamic analysis reveals bifurcation patterns which have not be seen in the Gray and Scott CSTR.

- The parameter boundaries at which the degeneracies $\mathbf{F_1}$, $\mathbf{H_{01}}$, $\mathbf{H_{02}}$, $\mathbf{H_{10}}$, $\mathbf{H_{20}}$, and $\mathbf{H_{11}}$ occur, have been identified in the κ, β–parameter plane. Note that $\mathbf{F_1}$, $\mathbf{H_{01}}$, and $\mathbf{H_{10}}$ degeneracies are the highest Hopf degeneracies which have been found in the Gray and Scott CSTR. For both models, the maximum number of Hopf points in any bifurcation diagram is four, which is similar to the case of the first-order exothermic reaction with extraneous heat capacitance. The maximum number of Hopf points is two in the CSTR case.

- Examining the codimension-one singularities (F_1, H_{01}, H_{10}) has revealed more than forty dynamic patterns for each model around these degeneracies. The two-phase reactor is considerably richer in behaviour than the CSTR, where only eleven bifurcation diagrams have been identified.

- Another feature that is an extension of the Gray and Scott CSTR is the presence of the codimension-two degeneracies (H_{20}, H_{11}), which are used to define regions in the parameter κ, β–space where multiple periodic solutions and isolas of periodic branches can be found.

- Secondary bifurcations of periodic solutions such as saddle-node, transcritical and period-doubling, several of which have not been previously found with the CSTR, have been observed for both models.

- We have found that when coupling is through the autocatalyst, the model exhibits a secondary Hopf bifurcation which produces quasi-periodic solutions close to the F_2 points. We have shown that the higher order degeneracies F_2 and G_1 occur only in the second model. Therefore, the second model appears to be more analogous to the nonisothermal case (first-order exothermic reaction with extraneous capacitance) which exhibits secondary bifurcation leading to tori.

12.3 Chaotic Behaviour

- There are no techniques available for determining a priori the regions of parameter values in which chaos is possible. However, we have found that identifying the period-doubling loci is the most important step in discovering the chaotic regimes. In the second model, period-doubling points exist in the vicinity of G_1 degeneracies.

- We have shown that both models are capable of displaying chaotic behaviour. Chaos, with periodic windows, can be found by varying the bifurcation parameter, θ. Calculations of Lyapunov exponents have proved very useful in certifying chaotic regions. Other tests such as time series, Poincaré maps and spectral analysis are also used to confirm regions of chaotic motions.

- The transitions to and from chaos occur either by period-doubling (Model I and Model II) or through a process involving intermittency (Model II). Both models have shown the kind of return map that indicates a qualitatively equivalent structure to difference equations with quadratic transformations.

12.4 Conclusions

- Despite their simplicity, these models support a full range of complex dynamic behaviour. Steady-state multiplicity, unique and multiple periodic solutions, quasiperiodicity, period-doubling and chaos have all been realized. Numerical evidence demonstrates that these three-variable schemes are among the simplest chemical systems with autocatalytic mechanisms that produce chaotic behaviour.

- An autocatalytic feedback element is essential to the generation of interesting phenomena in this type of isothermal system. In this work, we have shown that merely adding the linear exchange of reactant or autocatalyst between the two compartments provides a third variable to the Gray and Scott CSTR, that is sufficient to provoke the complete range of periodic and aperiodic behaviour.

- These isothermal models, in which one of the components plays the autocatalytic role, have many of the properties previously observed for nonisothermal systems, where the autocatalytic element is the heat. Thus, isothermal systems, which have fewer parameters and simpler nonlinear equations, can be used to infer possible types of behaviour of nonisothermal systems. The second model (coupling through the autocatalyst) is found capable of displaying the same types of Hopf degeneracies as the case of first-order exothermic reaction in a CSTR with extraneous thermal capacitance and of avoiding the extreme stiffness of the nonisothermal case.

• We have demonstrated the usefulness of the singularity theory, Lyapunov-Schmidt method and normal form method for studying the dynamic behaviour of chemical systems. They are used for determining the parameter values at which singularities occur and for identifying all varieties of dynamic behaviour in the regions of parameter space adjacent to the singular points.

• Three-variable autocatalytors are attractive models for developing a better understanding of the sources of complex oscillations and chaos in chemical systems.

BIBLIOGRAPHY

"A + 2B $\leftrightarrow$ 3B, B $\leftrightarrow$ C in a Continuous-flow Stirred-tank Reactor", *Proc. Roy. Soc.* **A 411**, 193-206, 1987.

Aluko, M., and H. Chang, "PEFLOQ: An Algorithm for the Bifurcational Analysis of Periodic Solutions of Autonomous Systems", *Computers and Chemical Engineering*, Vol. 8, No. 6, 355-365, 1984.

Argoul, F., A. Arneodo, and P. Richetti, "Experimental Evidence for Homoclinic Chaos in the Belosv-Zhabotinskii Reaction", *Physica Letters A*, **120**, No. 6, 269-276, 1987.

Argoul, F., A. Arneodo, P. Richetti, and J.C. Roux, "Chemical Chaos: From Hint to Confirmation", *Acc. Chem. Res.*, **20**, 436-442, 1987.

Aris, R., and N.R. Amundson, "An Analysis of a Chemical Reactor Stability and Control I, II, III", *Chem. Eng. Sci.*, **7**, 121-155, 1958.

Aris, R., "Forced Oscillations of Chemical Reactors" in *Spatial Inhomogeneities and Transient Behaviour in Chemical Kinetics*, Ed. P. Gray, G. Nicolis, F. Baras, P. Borckmans, and S.K. Scott, Manchester, University Press, 1990.

Aris, R., "The Mathematical Background of Chemical Reactor Analysis, II. The Stirred Tank", in *Reacting Flows: Combustion and Chemical Reactors*, AMS/SIAM Lectures in Applied Mathematics, **24**, 75-107. 1986.

Bailey, J.E., "Periodic Phenomena in Chemical Reactor Theory," In *Chemical Reactor Theory* (Ed. L. Lapidus, N.R. Amundson), Prentice-Hall Englewood Cliffs, NJ, 1977.

Baker, J.L., and J.P. Gollub, *Chaotic Dynamics: An Introduction*, Cambridge University Press, Cambridge, 1990.

Balakotaiah, V., and D. Luss, "Analysis of Multiplicity Patterns of a CSTR", *Chem. Eng. Commun.*, **13**, 111-132, 1981.

Balakotaiah, V., and D. Luss, "Dependence of the Steady States of the CSTR on the Residence Time", *Chem. Eng. Sci.*, **38**, 1709-29, 1983.

Balakotaiah, V., and D. Luss, "Structure of the Steady State Solutions of Lumped Parameter Chemically Reacting System", *Chem. Eng. Sci.*, **37**, 1611-23, 1982.

Balakotaiah, V., "On the Steady-state Behaviour of the Autocatalytor Model".

Bilous, O., and N.R. Amundson, "Chemical Reactor Stability and Sensitivity", *A.I.Ch.E. J*, **1**, 513-521, 1955e.

Boissonade, J., "Aspects Théoriques de la Douple Oscillations dans les Systéms Dissipatifs Chimiques", *J. Chim. Phys.*, **73**, 540-544, 1976.

Cordonier, G.A., L.D. Schmidt, and R. Aris, "Forced Oscillations of Chemical Reactors with Multiple Steady States", *Chem. Eng. Sci.*, **45**, 1659-1675, 1990.

D'Anna, H., P.G. Lignola, and S.K. Scott, "The Application of Singularity Theory to Isothermal Autocatalytic Systems: The Elementary Scheme $A + mB \rightarrow (m + 1) B$", *Proc. R. Soc. Lond.*, **A403**, 341-363, 1986.

Degan, H., and L.F. Olsen, "Bistability, Oscillations and Chaos in an Enzyme Reactions", *Annuals New York Academy of Science*, 1977.

Doedel, E.J., and J.P. Kerevez, "AUTO: Software for Continuation and Bifurcation Problems in Ordinary Differential Equations", Caltech, Pasadena, 1986.

Doedel, E.J., and J.P. Kerevez, "Software for Continuation Problems in Ordinary Differential Equations with Applications", Caltech, Pasadena, 1981.

Doedel, E.J., and R.H. Heinmann, "Numerical Computations of Periodic Solutions Branches and Oscillatory Dynamics of the Stirred Tank Reactor with A $\rightarrow$ B $\rightarrow$ C Reactions", *Chem. Eng. Sci.*, **38**, 1493-1499, 1983.

Farr, W.W., and R. Aris, " 'Yet Who Would Have Thought the Old Man to have had so much Blood in Him' - Reflections on the Multiplicity of Steady States of the Stirred Tank Reactor", *Chem. Eng. Sci.*, **41**, 1385-1402, 1986.

Farr, W.W., "Mathematical Modelling: Dynamics and Multiplicity", University of Minnesota, 1986.

Field, R.J., and M. Burger, (eds.), "Oscillations and Travelling Waves in Chemical Systems", Wiley-Interscience, New York, 1985.

Golubitsky, M., and B.L. Keyfitz, "A Qualitative Study of the Steady-state Solutions for a Continuous Flow Stirred Tank Reactor", *SIAM J. Math. Anal.*, **11**, 316-339, 1980.

Golubitsky, M., and W.F. Langford, "Classifications and Unfoldings of Degenerate Hopf Bifurcations", *J. Diff. Equations*, **41**, 375-415, 1981.

Golubitsky, M., and D. Schaeffer, *Singularities and Groups in Bifurcation Theory*, Vol. 1, Springer-Verlag, New York, 1985.

Golubitsky, M., I. Stewart, and D.G. Schaeffer, *Singularities and Groups in Bifurcation Theory,* Vol. 2, Springer Verlag, New York, 1988.

Gray, P., and S.K. Scott, "Archtypal Response Patterns of Open Chemical Systems with Two Components", *Philo. Trans. R. Soc.*, **A332**, 69-87, 1990b.

Gray, P., and S.K. Scott, "Autocatalytic Reactions in the Isothermal, Continuous Stirred Tank Reactor. Isola and Other Forms of Multiplicity", *Chem. Eng. Sci.*, **38**, 29-43, 1983.

Gray, P., and S.K. Scott, "Autocatalytic Reactions in the Isothermal, Continuous Stirred Tank Reactor. Oscillations and Instabilities in the System $A + 2B \rightarrow 3B; B \rightarrow C$", *Chem. Eng. Sci.,* **39**, 1087-97, 1984.

Gray, P., and S.K. Scott, *Chemical Oscillations and Instabilities: Nonlinear Chemical Kinetics*, Oxford University Press, 1990a.

Gray, P., and S.R. Kay, "Modelling Complex Oscillatory Behaviour by Simple Chemical Schemes", *J. Phys. Chem.* **94**, 3304-3308, 1990.

Gray, P., S.K. Scott, and R. Aris, "Modelling Cubic Autocatalysis by Successive Bimolecular Steps", *Chem. Eng. Sci.* **43**, 207, 1988.

Guckenheimer, J., and Holmes, P., *Nonlinear Oscillations, Dynamical Systems and Bifurcation of Vector Fields*, Springer, New York, 1983.

Hilborn, R.C., *Chaos and Nonlinear Dynamics*: *An Introduction for Scientists and Engineers*, Oxford University Press, New York, 1994.

Hopf, E., "Abzweigung Einer Periodschen Losung von Einer Stationaren Losung Eines Differential System", *Ber. Math. Phys. Kl. Sachs. Acad. Wiss. Leipzig*, **94**, 1-22, 1942.

Hudson, J.I., and O.E. Rossler, "Chaos in Simple Three and Four Variable System", in *Modelling of Patterns in Space and Time,* (Ed. W. Jager and J.D. Murray), Springer, Berlin, 1984.

Jorgenson, D.V., and R. Aris, "On the Dynamics of a Stirred Tank with Consecutive Reactions", *Chem. Eng. Sci.*, **39**, 1741-1752, 1983.

Jorgenson, D.V., and W.W. Farr, and R. Aris, "More on the Dynamics of the Stirred Tank with Consecutive Reactions", *Chem. Eng. Sci.*, **39**, 1741-1752, 1984.

Kay, S.R., S.K. Scott, and P.G. and Lignola, "The Applications of Singularity Theory to Isothermal Autocatalytic Reactions: The Influence of Uncatalyzed Reactions", *Proc. R. Soc.*, **A409**, 433-448, 1987.

Keener, J.P., "Secondary Bifurcation and Multiple Eigenvalues", *SIAM J. Appl. Math*, **37**, 330-49, 1979.

Kevrekidis, I.G., L.D. Schmidt, and R. Aris, "Forcing an Entire Bifurcation Diagram: Case Studies in Chemical Oscillators", *Physica*, **23D**, 391-395, 1986.

Kevrekidis, I.G., L.D. Schmidt, and R. Aris, "Some Common Features of Periodically Forced Reacting Systems", *Chem. Eng. Sci.*, **41**, 1263-1276, 1986.

Langford, W.F., "A Review of Interactions of Hopf and Steady-state Bifurcations", in *Nonlinear Dynamics and Turbulence*, (Ed. G.I. Barenblatt, G. Ioose, and D. Joseph), 215-237, Pitman, Boston, 1983.

Langford, W.F., "Periodic and Steady-state Mode Interactions Lead to Tori", *SIAM J. Appl. Math.*, **37**, 22-48, 1979.

Leach, J.A., J.H. Merkin, and S.K. Scott, "An Analysis of a Two-cell Coupled Nonlinear Chemical Oscillator", *Dynamics and Stability of Systems*, **6**, 341-365, 1991.

Li, T.Y., and J.A. Yorke, "Period-Three Implies Chaos", *Amer. Math. Monthly*, **82**, 985-992, 1975.

Liljenroth, F.G., "Starting and Stability Phenomena of Ammonia Oxidation and Similar Reactions", *Chem. Metall. Engng.*, **19**, 287-297, 1918.

Lin, K.F., "Concentration Multiplicity and Stability for Autocatalytic Reactions in a CSTR", *The Cand. J. Chem. Eng.*, **57**, 467-480, 1979.

Lin, K.F., "Multiplicity, Stability and Dynamics for Isothermal Autocatalytic Reactions in a CSTR", *Chem. Eng. Sci.*, **36**, 1447-1452, 1981.

Lotka, A.J., "Contribution to the Theory of Periodic Reactions", *J. Phys. Chem.-USA*, 271-274, 1910.

Lyberatos, G., B. Kuzsta, and J.E. Bailey, "Versal Matrix Families, Normal Forms and Higher Order Bifurcations in the Dynamics of Chemical Systems", *Chem. Eng. Sci.*, **40**, 1177-1189, 1985.

Lynch, D.T., "Chaotic Behaviour of Reaction Systems" Parallel Cubic Autocatalytor", *Chem. Eng. Sci.*, **47**, 347-355, 1992a.

Lynch, D.T., "Chaotic Behaviour of Reaction Systems: Consecutive Quadratic/Cubic Autocatalysis with Intermediates", *Chem. Eng. Sci.*, **48**, 2103-2108, 1993.

Lynch, D.T., "Chaotic Behaviour of Reaction Systems: Mixed Cubic and Quadratic Autocatalysis", *Chem. Eng. Sci.*, **47**, 4435-4444, 1992b.

Lynch, D.T., T.D. Rogers, and S.E. Wanke, "Chaos in a Continuous Stirred Tank Reactor", *Math Modelling*, 3, 103-116, 1982.

Marek, M. and I. Schreiber, *Chaotic Behaviour of Deterministic Dissipative Systems*, Cambridge University Press, Cambridge, 1991.

Marsden, J.E., and M. McCracken, *The Hopf Bifurcation and Its Applications*, Springer, New York, 1976.

McKarnin, M.A., L.D. Schmidt, R. Aris, "Responses of Nonlinear Oscillators to Forced Oscillations: Three Chemical Reactor Case Studies", *Chem. Eng. Sci.*, **43**, 2833-2844, 1988.

Moon, C.F., *Chaotic and Fractal Dynamics: An Introduction for Applied Scientists and Engineers*, Wiley, New York, 1992.

Nayfeh, A.H., *Method of Normal Forms,* Wiley, New York, 1993.

Newhouse, S., D. Ruella, and F. Takens, "Occurrence of Strange Axiom A Attractor New Quasiperiodic Flows on T_m, m = 3 or more", *Commun. Math. Phys.*, **64**, 35, 1978.

Parker, T.S., and L.O. Chau, *Practical Numerical Algorithms for Chaotic Systems*, Springer-Verlag, New York, 1986.

Peng, B., S.K. Scott, and K. Showalter, "Period-doubling and Chaos in a Three-variable Autocatalytor", *J. Phys. Chem.*, **94**, 5243-5246, 1990.

Pismen, L.M., "Dynamics of Lumped Chemically Reacting Systems near Singular Bifurcation Points - II Almost Hamiltonian Dynamics", *Chem. Eng. Sci.*, **40**, 905-916, 1985.

Planeaux, J.B., and K.F. Jensen, "Bifurcation Phenomena in a CSTR Dynamics: A System with Extraneous Thermal Capacitance", *Chem. Eng. Sci.*, **41**, 1497-1523, 1986.

Pomeau, Y., and P. Manneville, "Intermittent Transition to Turbulance in Dissipative Systems", *Commun. Math. Phys.*, **74**, 189-197, 1980.

Salnikov, I.Ye., "Thermokinetic Model of a Homogeneous Periodic Reaction", *Dokl. Akad. Nauk.* **60**, 405-8, 1948.

Salnikov, I.Ye., "Thermokinetic Model of a Homogeneous Periodic Reaction", *Zh. Fiz. Khim.* **23**, 258-60, 1949.

Scott, S.K., and W.W. Farr, "Dynamic Fine Structure in the Cubic Autocatalytor", *Chem. Eng. Sci.* **43**, 1708-10, 1988.

Scott, S.K., *Chemical Chaos*, Clarendon Press, Oxford, 1991.

Seyedel, R., *From Equilibrium to Chaos*, Elsevier, New York, 1988.

Shil'nikov, L.P., "A Contribution to the Problem of the Structure of an Extended Neighborhood of a Rough Equilibrium State of Saddle-focus Type", *Math. USSR Sbornik*, **10**, 91-102, 1970.

Shimada, I., and T. Nagashima, "A Numerical Approach to Ergodic Problem of Dissipative Dynamical Systems", *Progress of Theoretical Physics*, V61, No. 6, 1605-1616, 1979.

Troger, H., and A. Steindl, *Nonlinear Stability and Bifurcation Theory: An Introduction for Engineers and Applied Scientists*, Springer-Verlag Wien, New York, 1991.

Troy, W.C., "Mathematical Analysis of the Oregonator Model of the Belousov-Zhabotinskii Reaction", in *Oscillations and Travelling Waves in Chemical Systems* (Ed. R.J. Field and M. Burger), Wiley, New York, 1985.

Tyson, J.J., "The Belousov-Zhabotinskii Reaction", *Lect. Notes. Biomath.*, **10**, 1976.

Uppal, A., W.H. Ray, and A.B. Poore, "The Classification of the Dynamical Behaviour of Continuous Stirred Tank Reactors", *Chem. Eng. Sci*, **31**, 205-214, 1976.

Van Heerden, G., "Autothermic Processes", *Ind. Engng. Chem.*, **45**, 1242-1247, 1953.

Wiggins, S., *Global Bifurcations and Chaos: Analytical Methods*, Springer-Verlag, New York, Heidlberg, Berlin, 1988.

Wiggins, S., *Introduction to Applied Nonlinear Dynamical Systems and Chaos*, Springer-Verlag, New York, Heidlberg, Berlin, 1990.

Wolf, A., "Quantifying Chaos and Lyapunov Exponents", *Nonlinear Scientific Theory and Applications*, Ed. A.V. Holden, Manchester, University Press, Manchester, 1984.

INDEX

This is an index of all the figures that have numerical values of the residuary parameters, σ, λ and ρ arranged lexically in order of increasing magnitude. When one of these parameters is used on a coordinate axis, that fact is denoted by C since it is being used as a continuous variable (it would have made the index too cumbersome to have given full details of the range); when one is being used as a label on a series of curves, this fact is denoted by D, since the parameter then has discrete values. If the entry is left blank the parameter is irrelevant, as ρ is to the steady states of Model I. The number of branch set figures (β, κ-planes divided into regions associated with particular bifurcation diagrams) is recorded under BS, of bifurcation diagrams (θ, x-planes with dispositions of the steady states and periodic solutions) appears under BD and of model trajectories under MT. The trajectories are distinguished by a following number in parentheses: (1) for a time series; (2) for a projection on a plane; (3) for a representation in x, y, z-space; (4) for a combination of projections on the coordinate planes with another element; (6) for figures with six elements. All other plots come under 'Other' with the number of them in the first column And the abcissa and ordinate in the second; Poincaré, first return maps, next maxima and Lyapounov exponent curves are denoted by π, ϕ, μ, and ε respectively. For any purely schematic figures or figures where the axes are not numerated, the number of figures is enclosed in square brackets, i.e., [.].

322

σ	λ	ρ	BS	BD	MT	Oth er		Model	Figure	Page
0	0	-	1					I	4.07	63
0.1	20	20	1					II	8.15	206
0.1	20	20		2				II	8.17	208
0.1	20	20			1(4)			II	8.18	209
0.1	20	20		4				II	8.19	210
0.1	20	20			2(1)			II'	8.22	219
0.1	20	20				1	σ,κ	II	8.23	220
0.1	100	2	2					II	11.01	277
0.1	100	2		6				II	11.02	278
0.1	100	2		3				II	11.07	284
0.1	100	2			4(3)			II	11.08	285
0.1	100	2				1	ε	II	11.09	287
0.1	100	2			1(6)			II	11.1	289
0.1	100	2			1(6)			II	11.11	289
0.1	100	2			1(6)			II	11.12	290
0.1	100	2			1(6)			II	11.13	290
0.1	100	2			1(6)			II	11.14	291
0.1	100	2			1(6)			II	11.15	291
0.1	100	2			1(6)			II	11.16	292
0.1	100	2			1(6)			II	11.17	292
0.1	100	2			1(6)			II	11.18	293
0.1	100	2			2(1,3)			II	11.19	293
0.1	100	2			1(1)			II	11.2	295
0.1	100	2			1(1)			II	11.21	295
0.1	100	2			1(1)			II	11.22	296
0.1	100	2			1(1)			II	11.23	296
0.1	100	2			1(1)			II	11.24	297
0.1	100	2			1(1)			II	11.25	297
0.1	100	2			1(2)			II	11.26	298
0.1	100	2				2	π	II	11.27	299
0.1	100	2				1	π	II	11.28	300

σ	λ	ρ	BS	BD	MT	Oth er		Model	Figure	Page
0.1	100	2				2	φ	II	11.29	300
0.1	100	2				1	ε	II	11.3	302
0.1	100	2			4(3)			II	11.31	303
0.1	100	2			1(3)			II	11.32	304
0.1	100	2				1	π	II	11.33	305
0.2	C	D				1	κ,λ	I	7.21	152
0.2	C	D				1	β,λ	I	7.21	152
0.5	1	1	1					II	8.04	191
0.5	4	5	1	[16]				I	7.16	147
0.5	9	5	1					I	7.31	167
0.5	9	5	1					I	7.42	180
0.8	100	2		1				II	11.05	282
0.8	100	2		1	5(1)			II	11.06	283
1	0.04	5				1	κ,θ	I	7.01	126
1	0.04	5		[4]				I	7.02	127
1	0.04	5		1				I	7.03	127
1	0.04	5		1	2(1)			I	7.04	128
1	0.04	5		1	1(1)			I	7.05	128
1	0.04	5	1	[14]				I	7.13	144
1	0.04	5	1					I	7.29	165
1	0.04	5	1					I	7.42	180
1	1	1	1					II	5.01	76
1	1	1	1					II	5.02	79
1	1	1	1	4				II	5.03	79
1	1	1	1					II	8.02	187
1	1	1		14				II	8.03	189
1	1	2	1					II	8.29	224
1	1	2	1	7				II	8.3	225
1	1	-	1					I	4.01	55
1	1	-	1					I	4.02	58
1	1	-	1	[4]				I	4.03	59

σ	λ	ρ	BS	BD	MT	Oth er		Model	Figure	Page
1	1	D	1					II	5.11	87
1	4	1	1	2				I	7.06	131
1	4	1		[4]		3	κ.θ	I	7.07	132
1	4	1	1					I	7.08	138
1	4	1		11				I	7.09	139
1	4	1	1					I	7.24	155
1	4	1		14				I	7.25	156
1	4	1	1					I	7.34	172
1	4	5	1	[12]				I	7.1	141
1	4	5	1					I	7.26	158
1	4	5		20				I	7.27	159
1	4	5	1					I	7.34	172
1	4	5			3(2,1)			I	7.35	173
1	4	5	1					I	7.4	177
1	4	5		5				I	7.41	178
1	4	5	1					I	7.42	180
1	4	20	1	[12]				I	7.11	143
1	4	20	1					I	7.19	150
1	4	20				2	κ,θ	I	7.2	151
1	4	20	1					I	7.28	164
1	4	20	1					I	7.34	13
1	4	20	1					I	7.36	174
1	4	20	1					I	7.4	177
1	4	20	1					I	7.42	180
1	4	20	2					I	7.43	181
1	4	20		1	1(1)			I	7.44	182
1	4	D	1					I	7.12	143
1	10	1	1	6				II	5.16	94
1	400	5	2	[16]				I	7.14	145
1	400	5	1					I	7.3	166
1	C	D				1	κ,λ	I	7.22	152

σ	λ	ρ	BS	BD	MT	Oth er		Model	Figure	Page
1	C	D				1	β,λ	I	7.22	152
1	D	1	1					II	5.1	86
1	D	5	1					I	7.15	146
1	D	-	1					I	4.06	61
2	2.25	4		4				I	7.38	175
2	2.25	5	1	[9]				I	7.17	148
2	2.25	5	1					I	7.32	168
2	2.25	5	1					I	7.34	172
2	2.25	5	1					I	7.37	175
2	2.25	5		1	1(1)			I	7.39	176
2	225	5		1				I	10.02	244
2	225	5			5(1)			I	10.03	245
2	225	20		3				I	10.04	247
2	225	20				1	ε	I	10.05	249
2	225	20			16(4)			I	10.06	251
2	225	20			16(3)			I	10.07	260
2	225	20			4(3)			I	10.08	263
2	225	20				16	π	I	10.1	266
2	225	20				3	π	I	10.11	268
2	225	20				6	μ	I	10.12	270
2	225	20				16	ϖ	I	10.13	272
5	0.1	1	1					II	8.09	196
5	1	0.1	1					II	8.07	194
5	1	1		1				II	8.01	185
5	1	1	1					II	8.05	192
5	1	1		[22]				II	8.06	193
5	1	1	1					II	8.24	220
5	1	1				2	κ,θ	II	8.25	221
5	1	1	1	3				II	8.31	226
5	1	10	1					II	8.08	195
5	10	1	1					II	8.1	197

σ	λ	ρ	BS	BD	MT	Oth er		Model	Figure	Page
5	10	1		25				II	8.11	198
5	C	D				1	κ.λ	I	7.23	153
5	C	D				1	β,λ	I	7.23	153
8	7	1	1					II	8.12	200
8	9	1	1					II	8.13	201
8	9	1		[40]				II	8.14	203
8	9	1				3	κ,θ	II	8.26	222
9	C	D				1	λ,β	II	5.07	84
9	C	D				1	λ,κ	II	5.07	84
10	0.303	-	1					I	4.09	69
10	0.303	-		6				I	4.1	70
17.8	25	1		3				II	5.15	91
C	100	D				3	θ,σ	II	11.04	281
C	C	D				1	σ,λ	II	5.14	90
C	D	1				1	κ,σ	II	8.27	223
C	D	2				3	θ,σ	II	11.03	279
C	D	4				1	κ,σ	II	8.28	223
C						1	θ,σ	I	10.01	243
D	1	1	2					II	5.09	86
D	1	-	2					I	4.05	61
D	4	5	1					I	7.18	149
D	C	1				3	λ,β	II	5.04	81
D	C	1				1	λ,β	II	5.06	83
D	C	1				1	λ,κ	II	5.06	83
D	C	-				2	λ,β	I	4.04	59
D	C	-				2	λ,β	I	4.08	69
D	C	D				1	λ,β	II	5.13	90
D	D	1	3					II	5.05	82
D	D	C				3	ρ.κ	II	8.15	206
D	D	C				3	ρ,β	II	8.16	207
	C					1	θ,λ	I	10.01	243

σ	λ	ρ	BS	BD	MT	Oth er		Model	Figure	Page
		C				1	θ,ρ	I	10.01	243
				[41]				I	7.33	170